RADIO
MONITORING
THE HOW-TO GUIDE

"Never has so much useful information about getting started in so many aspects of radio monitoring been packed into one book . . . . This complete how-to guide should grace the shelf of anyone in the radio hobby."
—Association of North American Radio Clubs

T.J. "SKIP" AREY
N2EI

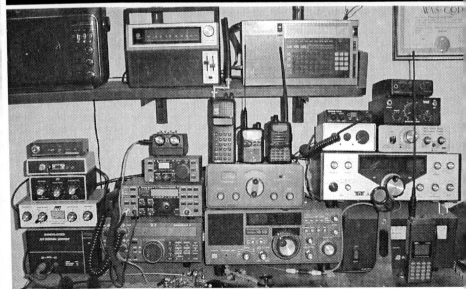

PALADIN PRESS · BOULDER, COLORADO

Radio Monitoring: The How-To Guide
by T.J. "Skip" Arey, N2EI

Copyright © 1997, 2003 by T.J. "Skip" Arey, N2EI

ISBN 10: 1-58160-405-X
ISBN 13: 978-0-87364-405-4
Printed in the United States of America

Published by Paladin Press, a division of
Paladin Enterprises, Inc.
Gunbarrel Tech Center
7077 Winchester Circle
Boulder, Colorado 80301 USA
+1.303.443.7250

Direct inquiries and/or orders to the above address.

Visit our Web site at www.paladin-press.com

DEDICATIONS
AND DEBTS OF GRATITUDE

This book is dedicated to my wife Regina and to my sons Brendan and Patrick for putting up with me as I spent so much time in the "writing mode" over the last few months. I guess I have some chores to do around the house.

I also need to thank a few folks for being there along the way:

Robert Anson Heinlein for teaching me about many things.
Hank Bennett for the beginner's book that got me started a long time ago.
Bill Oliver for convincing me I could write about radio.
Bob and Judy Grove who gave me the chance to get the word out.
Larry Miller and Rachel Baughn who edited me well over the years.
Linton Vandiver for teaching me how to think beyond 2000 words.
Bill Cheek for keeping the Internet interesting and for the graphics help.
Henry Eisenson for the kind thoughts, support, and help with my hyphens.
Ralph Brandi for the Web List.
My radio monitoring compatriots, *THE SCANNER SCUM*:

Dave Marshall N8OAY		Jon Cohen WB2KKS
John McColman N4RVR		Mark Meece N8ICW
Tom Swisher N8GQK		John Vodenik WB9AUJ
Mike Eilers		Bill Cole
Lynn Burke	Ed Muro	Don Schimmel

And finally thanks to all the radio monitoring clubs, large and small, and also to everyone who has ever given a hand and a few words of encouragement to a beginning radio monitor.

TECHNICAL INFORMATION

This book was written on a Comtrade 60 MHz Pentium PC and a Samsung NoteMaster 386S/25 Laptop PC. The word processing software was Lotus Ami Pro 3.0 for Windows (with a Microsoft Word export filter). Reference Software's Grammatik 5 for Windows took care of all those things I missed in Mr. Gallagher's English classes. Much of the business conducted around getting this project to fly occurred by way of the Internet using Netscape 2.01. Graphics were created enhanced, redrawn, and/or processed by Bill Cheek using an HP ScanJet 3P page scanner, Microsoft Draw, Microsoft Word 7.0, Aldus PhotoStyler, Softkey PhotoFinish, and JASC PaintShopPro under Microsoft Windows 95™.

CONTENTS

Preface to the Second Printing

When I first wrote *Radio Monitoring: The How-To Guide*, I did so with the intention of presenting the information in a way that would keep it useful through future printings. The bulk of the information in this book is about how to listen to radio signals, regardless of the frequency in question or the technology brought to bear on the task. Presenting these skills independent of the signal and hardware aspects of radio monitoring has been what has made this book both unique and well-received by the radio hobby community since it was first released. Where to listen is always changing. What you use to listen is always changing. The basic skills that make a radio monitor successful have applied universally since the days of Marconi, Armstrong, and Fessenden.

All that being said, there are a few concepts that would benefit from a bit of shoring up.

At the time of the first printing, radio monitoring hobbyists were just beginning to see the benefits of the Internet and the World Wide Web as tools for gathering and exchanging information. Type "shortwave" into your Internet search engine of choice and you will find literally hundreds of references beyond the basic sites presented in this book's original listings.

In the area of shortwave broadcast stations, many signals can now be heard over the Internet. Some people initially saw this as at least a distraction or possibly even signaling the death of the radio monitoring hobby. This has proven not to be the case. Online monitoring of radio signals has proven to be a useful adjunct to on-the-air listening. Further, the ability to hear VHF/UHF public safety signals from distances well beyond local signal limits opens up an entire new world of monitoring.

Digital Signal Processing (DSP) has presented many improvements for the radio monitor. DSP receiver designs were once limited to nearly unobtainable commercial and military receiver designs costing many thousands of dollars. Now DSP circuitry and its benefits can be found in quality receiving equipment in almost any price range. Anyone in the market for a new receiver will want to pay close attention to the improved selectivity that DSP provides.

The ability to interface a receiver (and transmitter) with a personal computer continues to be an area of exciting experimentation. Using the PC's sound card to process non-voice communication brings RTTY, Packet Radio, fax, CW, and other signals to the radio monitor for little more than the cost of a couple of patch cords and some software easily found on the Internet. Amateur radio enthusiasts have taken this a step further and are exploring digital transmission with such technologies as PSK31. As this book points out repeatedly, a curious mind will always find new and interesting things to monitor.

Digitally encoded and managed signals have changed the way public service frequencies are being monitored. "Trunked" radio systems have come on the scene in most major metropolitan areas as a way of making more efficient use of the radio spectrum. Before purchasing a UHF/VHF scanning receiver for any given area, the user will want to have a clear understanding of local signal systems and protocols. Scanning receivers are no longer a "one size fits all" product. The good news is that receivers are always coming on line to address these unique situations, as the public always seeks to remain informed about the organizations that serve it.

I would be remiss if I did not talk about the perception, on the part of the general public, that the world has changed since the events of September 11, 2001. The shift of world events that started on that terrible day, while shocking, was probably not nearly as surprising to the experienced radio monitor. Radio monitoring hobbyists are the most informed people on the planet. They know that the world is always changing, always changing rapidly, often changing in unexpected ways, and, sadly, not always changing for the better. While no one can accurately predict what will happen in the future, people who take the time to "listen to the world" will always have the edge.

Since the first printing of this book, the radio monitoring hobby lost one of its greatest proponents, Bill Cheek. Bill had a lot to say and a lot to teach. Much of his wisdom and information can still be found in his books, and I commend them to anyone who wants to go further in this hobby.

My close friend Jon Cohen WB2KKS rightly pointed out that on page 133 I used the term "Longwire" when I was actually describing an "End Fed Wire." My only glaring error. I told you this was a good book! Have fun.

T.J. "SKIP" AREY, N2EI
January 2003

Introduction

Everybody needs a hobby. It is all but essential in this rapidly moving world, to have an activity that affords a moment to relax from the normal affairs of the day. Some people play golf, softball, or basketball. Some paint, sculpt, or carve. Still others camp, climb mountains, or surf. Most hobbies provide their relaxation and recreation by separating the hobbyist from the world at large.

I will show you a hobby that provides all the traditional benefits and still allows you to engage the world around you in a unique and exciting way. My hobby of choice is *Radio Monitoring*. Radio monitoring is an exploration and investigation of the radio frequency spectrum. Radio monitoring is tuning between and beyond the three or four broadcast stations you listen to in your car on the way to work. Radio monitoring is hearing things that weren't meant for your ears, but which provide a great deal of excitement anyway.

Virtual reality has become an overused term, describing personal experiences with computers. Virtual reality usually "tricks" the mind into feeling that it is somewhere that it isn't. The experienced Radio Monitor smiles knowingly when the term "virtual reality" is tossed in his or her direction. Radio has been a form of virtual reality since its discovery over a century ago. Virtual reality of radio opens the mind to experiences and knowledge from more distant horizons.

Have you ever dreamed of traveling the world in search of adventure? Most of us have at least daydreamed about exploring exotic places in distant lands. But few of us have the time or money to make regular world wide treks for excitement. We have to find our adventure closer to home. It is neither safe nor smart to chase police cars, fire engines, or emergency service vehicles. Maybe you are curious about how people are affected by a newsworthy event in another state. Your local news broadcast might not be as complete a picture as one from the location where things are actually happening.

Radio monitoring is a unique and affordable adventure. You can experience the events and culture of almost any place on our planet. Enter the radio monitor's world, a place of wonder and excitement by tuning into the virtual reality of the radio frequency spectrum.

Most people's experience with radio is limited to the standard AM and FM broadcast bands found on their car or home radios. Practically, this represents only a tiny portion of the radio frequencies available. There is a vast bandwidth of frequencies within easy and inexpensive reach that can provide countless hours of excitement, education, and enjoyment to anyone with a desire to know more about the world around them. Radio monitoring is both the hobby and the sport of the naturally curious person.

Radio monitoring brings the world to your doorstep. Many countries broadcast specifically for audiences in North America as well as other parts of the world. Would you be surprised to hear that some of these signals can even be heard using that standard AM/FM car radio? Still more signals from foreign lands abound in the vast bandwidth between those fairly small bands of frequencies that make up AM and FM. A large part of the area between the more familiar frequencies is called the Shortwave Band. These signals are available for less than the price of many electronic children's games. Since you opened this book, you probably have at least a mild curiosity about the world of radio monitoring. As you progress through these pages, you will learn how to gain access to the radio frequency spectrum and all its wonders and adventures. I will discuss the types of receivers you will need, the antennas and accessories that will improve your ability to bring the world home, and the book covers all of the major facets of the radio monitoring hobby including how to get into contact with the stations you have heard. You will also learn how to connect with other radio monitors to enhance the fun and enjoyment of your hobby.

Radio monitoring is a multifaceted hobby. This book will give you all of the basic information that will allow you to begin enjoying the world of radio monitoring. If a particular section of the book raises some questions for you, note them and read on because it is likely that what you are looking for is covered in another segment.

> **The worldwide shortwave radio listening population numbers in the hundreds of millions.**

Because radio monitoring is such a wide and varied hobby, you will need information to assist in investigating advanced or specialized areas. You will learn about the many clubs and publications that are available to guide you in the advancement of radio monitoring.

No single book has all the information there is to be shared about the radio hobby. This book is designed to be a practical guide for the beginning, listener the "first" word, not the "last." You will gain access to information that will resolve most of the confusion that can stand in the way of full enjoyment as you grow in knowledge of the hobby.

This book will also break down the myths that radio monitoring is expensive or can only be enjoyed by a person with a Ph.D. in Electronics. *Nothing could be farther from the truth!* Compared to other hobby pursuits and interests, radio monitoring is an affordable activity that can be enjoyed by the entire family. It is a hobby for young and old alike. It fits into most budgets, even if a paper route, allowance, or fixed income. This is a hobby where tenacity can beat pocketsful of money. Your skills and equipment can advance at a personal and financially comfortable pace, and you will have fun every step of the way. You already possess 90 percent of what it takes to be successful in this hobby before you buy any equipment. I will point you in the path from beginner to advanced.

Radio monitoring can be a lifelong pursuit. The world is always changing. There are always new signals to hear and new things to discover. If you take advantage of the information that follows, you will develop a sound foundation in the radio monitoring art that will increase your enjoyment through the years, no matter the changes.

In this book I will also try to help you enjoy your beginner's status. There is no need to compare your accomplishments with those of others unless that level of competition appeals to you. This is a hobby where even the most experienced still recall their earliest listening activities with great enthusiasm. You will learn to document your adventures so you can relive your experiences again and again.

In this book I will periodically remind you that radio monitoring is a *hobby*. It's supposed to be *fun*. It is not supposed to give you the same aggravation you get from the salt mines every day. You will discover a hobby that gives a great sense of accomplishment without driving you to total distraction. You will see that I take this subject and myself only as seriously as is needed to convey the point.

We will look primarily at the three most common forms of radio monitoring: **mediumwave** (the traditional AM broadcast band), **shortwave** (the primary frequencies for worldwide radio signals), and **VHF / UHF** (those signals most commonly monitored by a scanning receiver). Each of these differ enough to warrant separate consideration. We will also touch on other radio monitoring experiences into which a beginner might later want to branch.

> **The radio frequency spectrum covers nearly three thousand billion Hertz (Hz). The standard AM/FM car radio covers less than one thousandth of all these possible frequencies.**

I have been a radio monitor for more than 25 years. I've tracked down signals from "DC to Daylight" (a common term that suggests most of the RF spectrum). In that time I've monitored wars and rumors of wars. I've heard the rise and fall of nations. I've eavesdropped on the activities of spies and international criminals. I've enjoyed radio programming from every region of the United States and numerous foreign lands. I've tapped into the excitement in my own back yard on the public safety bands. I've never lost the excitement that comes from hearing a new signal. By way of radio monitoring I have traveled the planet, gotten to know the people of the world in unique and exciting ways. Any journey is more fun when it is shared. I welcome you along. We're going to have fun.

2

A brief history of radio

Most folks have heard the name Marconi and have a vague notion that he invented radio. It might be more appropriate to say that Guglielmo Marconi discovered how to exploit radio frequency energy. Like so many inventions, Marconi was only one among a number of tenacious experimenters who were looking into sending signals through the "ether." Other names with controversial credit with fathering radio are Dr. Mahlon Loomis, William H. Preece, and Nathan Stubblefield. Loomis conducted successful experiments in "receiving" transmitted electrical signals in 1865. Preece demonstrated a form of wireless communication in 1886. Stubblefield patented the *Vibrating Telegraph* in 1888, but the beginnings of radio actually go farther back.

Early experiments

The science behind the experimental efforts at the roots of radio can be traced to the early 1800s when Michael Faraday and Joseph Henry developed the theory that a current flowing through one wire could induce a current to flow in a separate wire. In 1820 Hans Christian Oersted demonstrated this principle of Electromagnetic Induction. It was the German physicist Heinrich Hertz who came up with the first reasonable explanation for what was going on, with his demonstration of the existence of electromagnetic waves in 1887. In 1892, French physicist Edouard Branly developed the first "receiver" of electromagnetic waves which he called a "Coherer." The experimenters of the world now had the science on which to build their ideas.

It was a young Marconi who followed these developments and created his first successful wireless system in 1895. After patenting his invention in England in 1896 he went on to pursue its commercial aspects by installing his system on ships and at shore stations. On December 12, 1901, Marconi successfully sent a signal from St. John's, Newfoundland to Poldhu, England. It was only the letter "S" sent in telegraphic code, but it was the first transatlantic broadcast of an electromagnetic signal. The world was amazed and interest in radio grew.

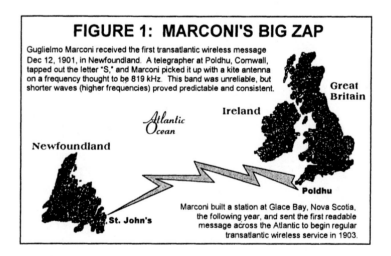

FIGURE 1: MARCONI'S BIG ZAP

Guglielmo Marconi received the first transatlantic wireless message Dec 12, 1901, in Newfoundland. A telegrapher at Poldhu, Cornwall, tapped out the letter "S," and Marconi picked it up with a kite antenna on a frequency thought to be 819 kHz. This band was unreliable, but shorter waves (higher frequencies) proved predictable and consistent.

Great Britain

Ireland

Atlantic Ocean

Newfoundland

Poldhu

St. John's

Marconi built a station at Glace Bay, Nova Scotia, the following year, and sent the first readable message across the Atlantic to begin regular transatlantic wireless service in 1903.

Early 1900's through World War I

In 1905 Sir Ambrose Fleming developed the Diode Electron Tube which permitted the detection of high frequency radio waves, a vast improvement over Branly's Coherer. Meanwhile, in 1906 Reginald

Aubrey Fessenden came up with a system that allowed transmission of the human voice instead of telegraph code. In 1907 Lee De Forest invented his "Audion," the first Triode Electron Tube to successfully amplify radio waves. In 1909 The Junior Wireless Club was formed in New York City. World War I put a halt to much amateur experimentation but military applications of radio technology further advanced the state of the art.

R. A. Fessenden was one of radio's pioneers

As you can see, things were happening fast. While experimenters developed improved radio techniques, more and more members of the general public were becoming excited by this new technology. Magazines such as *Popular Electricity* and *Radio* came into being, telling people about how radio was going to affect their lives and encouraging experimentation and use of radio. These magazines were also the source for the components that allowed people to build radio receiving equipment and this helped to further the hobby of radio monitoring. Many home tinkers also began to experiment with transmitting and the amateur radio hobby developed rapidly. In 1914, the American Radio Relay League was formed to support the growing number of amateur radio operators. Those people who constructed simple receivers to allow them to listen in on these strange and wonderful signals were having a ball. It became common practice to keep a log of the stations that they heard. Many listeners wrote to the stations requesting a confirmation card or letter and so the **QSL** (a telegrapher's abbreviation for *confirmation*) collecting aspect of the radio monitoring hobby was born. As you can see, radio monitoring grew up side by side with the technology of radio.

Edwin Armstrong

In 1918, Edwin Armstrong developed the superheterodyne receiver circuit. This made receiver tuning and adjustment the simple process that we are familiar with today whenever we use a modern radio receiver. Radio became accessible to the average individual, and thus commercial broadcasting was born.

Edwin H. Armstrong was the inventor of the basic circuits that made the standard AM/FM radio possible. He spent much of his life and fortune defending his patents. Tragically, he committed suicide in 1954. Can you imagine how far radio might have advanced had he continued his work?

Pioneering radio stations

In 1919 CFCX in Montreal, Quebec began commercial broadcasting, followed by WWJ in Detroit, MI and KDKA in Pittsburgh, PA in 1920. These stations' successes brought about a landslide of receiver purchases and new station construction. Everyone wanted to own a radio and anyone who had a product to sell (including radio receivers) or an idea to hawk wanted to build a transmitter. While these stations are often thought of as the first broadcasters, several early experimental stations became commercial broadcasters and can trace their roots back to earlier times. Two such stations are WQXR, New York, NY (now under the call letters WQEW) which began its existence as W2XR, and KQW, San Jose, CA, which is now KCBS, San Francisco, CA. This station is documented as having been in continuous operation since 1909.

Shortwave

As the understanding of radio technology and the effects of the radio frequency spectrum became more clear, stations began to experiment with the shortwave spectrum and the longer transmitting distances these higher frequencies brought about. Meanwhile radio was also growing in Europe and other parts of the world. For countries that did not have well developed traditional postal, telegraph, and telephone technologies, shortwave came as a blessing, allowing wide-ranging communications. One of the earliest radio monitoring clubs, the World Radio Reception Club of Pittsburgh, was formed in 1932.

Frequency modulation

In 1933 Edwin Armstrong developed yet another significant improvement to radio. **Frequency Modulation** (FM) allowed for radio transmission with reduced noise and interference. Development of FM and experimentation in the higher frequencies leading to the VHF and UHF regions began in earnest. 1933 was also the year that The National Radio Club was formed. This club was devoted to monitoring commercial AM broadcast band stations over long distances.

World War II

As with World War I, World War II brought about further advances in radio technology. The war years also signaled the birth of the Voice of America. Shortwave radio became a powerful tool for information (and propaganda) during the war. After the war was over, a new aspect of the radio hobby began. Many superior transmitters and receivers became available to the hobbyist by way of the surplus market. Military radio technology also filtered down into the commercial radio market. By the end of World War II, the world belonged to radio. Also, after the war, many radio applications beyond broadcasting began to develop. Public safety and business radio applications began to unfold.

Transistors

Things were going along swimmingly and then in 1948 Walter H. Brattain, John Bardeen, and William Shockley at Bell Laboratories invented the **transistor**. With Shockley's further improvements in 1952, this ushered in a whole new era of electronics and further improvements in radio receiver design. Vacuum tubes have an inherent problem due to their design. The heat generated by the tubes causes frequency drift in even the most sophisticated receivers. Circuits designed with transistors greatly improved upon this problem as well as many other aspects of receiver design. The late fifties found the country amazed by shirt pocket size radios. But folks hadn't seen anything yet.

Integrated circuits

In 1959 Robert Norton Noyce at Fairchild Semiconductor, and Jack Kirby at Texas Instruments, independently developed the first **integrated circuits** when they created networks of transistors on a single "chip." The miniaturization of electronics had begun. By the late sixties, entire radio receiver circuits could be placed on individual chips. Further developments in integrated circuitry that improved frequency synthesis and signal filtering came about as the result of the integrated circuit. The microprocessors that became available in the seventies brought us into the world of digitally-controlled radio electronics, including the modern scanning receiver.

Modern digital electronics has made it even more easy to enter and enjoy the radio monitoring hobby. A relatively modest-cost receiver of today contains high performance features that would have cost thousands of dollars only a few years ago.

Radio monitors of today carry on the tradition of curiosity and tenacity that the pioneers of radio demonstrated. We are no longer burdened with spending a month's wages to construct our own receivers from scratch to eke out a few dozen signals. We have the ability to use relatively inexpensive receivers to allow us to listen in on thousands of signals bringing every facet of the world into our own homes. Still, when you catch a rare station on the air after weeks of trying, it's fun to think of Marconi, Fessenden, Armstrong, and all the others who gave us the gift of radio.

3

Forms of listening

How many radios are in your house? I counted seven non-hobby radios in mine, but I am extremely radio-oriented, so I took a wider sample. Several neighbors not involved in the radio monitoring hobby revealed that their households possessed an average of four receivers, including those in their cars. Statistically, the United States averages two radio receivers per household, not counting the cars, so I guess you could say I live in an average neighborhood.

People listen to the radio more than they realize. A clock radio awakens us with the morning weather. On the way to school or work, we catch our favorite songs, the news, or the latest opinions of a talk show host. Likely a radio plays in the background at work. Back to the car radio for the commute home. During dinner and afterwards we tune some relaxing background music. Finally at the end of the day, we hit the sleep button on the clock radio for ten or fifteen minutes of music to lull us to sleep. Radio has always had a special advantage over television (and now over multimedia computer systems) because our bodies can do something else while our brains accept the input from a radio. Here might be a good place to show the difference between monitoring and just listening.

Radio monitors are usually more intent and organized when listening to their receivers. For the monitor, radio listening becomes the primary activity instead of the background information. A radio monitor is enjoying the programs that he or she hears, but the motivations are somewhat different. The casual radio listener probably monitors about half a dozen regular radio stations, depending on personal tastes. The radio monitor seeks out *stations never heard before*. The monitor collects stations and signals like other hobbyists might collect stamps, comic books, or bird sightings. Any station not heard is a subject of desire. The reasons why radio monitors go after these stations vary from hobbyist to hobbyist. For some folks it is simply the joy of adding each new station to the log. Others are trying to learn specific information about the world around them. Still others are attempting to understand how radio operates over various distances. These differences even play out in the terms that radio monitors use when referring to their practices.

As you become more involved in radio monitoring, you will come to notice that monitoring enthusiasts refer to themselves as **Listeners** or **DXers**.

Listeners

Listening is the more obvious of the two types of monitoring. For example, people who monitor the shortwave spectrum call themselves **SWLs**, standing for Shortwave Listener. Anyone who tunes the shortwave bands can accurately call themselves by this moniker. Anyone who monitors radio can be considered a Listener regardless of how they practice the hobby.

DX is a long-standing radio communications abbreviation for *distance*. Over the years, the term has taken on a broader meaning. Radio stations that were very far away from someone's reception point, such as those in other countries or beyond the normal expected signal range, became known as DX stations. The pursuit of these far-away signals became known as DXing and the person doing the pursuing became known as a DXer.

Over the years, consensus has developed the terms Listener and DXer so that they refer to two different styles of listening.

DXers

The Listener is usually the type of person who monitors various signals and stations for the sole purpose of enjoying the program content presented. This type of person may settle on a group of favorite stations or a particular kind of signal and remain happy for months or even years listening to these same stations.

The DXer, on the other hand, derives unique enjoyment from chasing after as many stations in as many places as possible. The goal of the skilled DXer is to catalog distant stations, regardless of program content. It is not unusual for the DXer who monitors the shortwave frequencies to seek out stations that broadcast in unknown languages, solely for the purpose of adding these new locations to the station log. Some mediumwave monitors specialize in tracking low-power signals from areas of the country that have little hope of getting their signal out beyond their local community. Some scanner enthusiasts take advantage of unusual band conditions to log signals over enormous distances never intended by the transmitting station. DXing is a highly challenging and fun aspect of the monitoring hobby.

Now let's explode a long standing myth you may encounter as you begin your radio monitoring experience. While some DXers and Listeners try very hard to prove they are different from one another, the fact is that most radio monitoring enthusiasts are really a little bit of both. Many people, especially those who are new to the radio monitoring experience, are enthralled by the ability to hear so many different signals. So it is natural to spend some time tuning through the radio frequency spectrum seeking out new and different signals, logging them and perhaps sending out requests for verification of reception. It is only after people become more experienced that they tend to follow either the Listener or DXer path as their main style of listening. Both are great fun, and many radio monitors dabble happily in both realms throughout their personal hobby history. In other words, as a beginner, don't get too hung up in trying to define yourself and your listening habits. There is so much out there for the beginner to experience that you can relax and enjoy whatever you are doing. You have nobody to impress but yourself.

Let's begin with a simple experiment in listening that will cost you absolutely nothing provided you are living in an average household with a radio or two in it. This experiment works best with any simple portable radio. Your only other "equipment" will be a pencil and paper. This experiment works best in the evening.

Many radio monitors collect "QSL" cards from distant stations

If the radio receiver you have access to is a typical AM/FM unit found in most homes, set it to the AM band and turn the tuning dial all the way down to the lowest frequency, usually listed as 540 or 54 on most receivers. Slowly tune the radio up through to the other end of the dial. Take some brief notes about what you're hearing. Don't get all wrapped up in logging any stations just yet. That will come later in the book when we discuss mediumwave listening specifically. All I want you to do at this point is make note of what you hear and how what you're hearing sounds. Obviously, every turn or so of the dial will bring up a different station. The local stations you are more familiar with will probably sound very strong without any interference or noise on their signal. But don't pay these stations any mind right now. If anything, they are more or less in the way for the purposes of this experiment. In between these strong locals you will begin to notice other, weaker signals. These signals may fade in and out or have a fluttering sound to them. When you hear one of these weak stations stop tuning and listen. These are signals that are far enough away from your local area to be affected by those conditions that impact on a radio signal going from the transmitter location to your receiver in your home. As you listen, you may begin to notice patterns to these effects. What you are beginning to learn about is the science of **propagation**. This is the study of how energy moves from one place to another through space. In this case, the energy we are concerned with is radio-frequency energy. As you move through your radio monitoring experiences you will learn how to turn information about radio-signal propagation into loggings of distant and unusual signals.

As you continue to tune through the AM band you will probably hear more than one signal on the same frequency, essentially interfering with each another. Stay on this frequency for a while and just listen to what you hear. Every now and then, it is likely that one of the signals will rise above the combined noise and be more easily understood. This, again, occurs in response to the effects of the propagation of the various signals on the frequency. But you've already discovered this phenomenon, haven't you?

What else is there to learn? Assuming the receiver is sitting before you on a table, physically turn the entire receiver 90 degrees in either direction. Notice any change? In most cases you will hear some of the signals weaken or even drop out of the range of hearing. You just made another important discovery. Radio frequency energy has directional properties. As you move forward in your experiences at radio monitoring you will further discover that this directionality can be made to occur at the transmitting site to get a signal to a particular place, and it can be used at a receiving location to assure reception of a particular signal. In this case, advantage is taken of the directional receiving properties of the antenna inside an AM receiver to "null" out signals that come to your receiver from specific directions. If you are looking for a cheap and dirty science fair project for yourself or a child, this simple form of **Radio Direction Finding** (RDF) has a lot of possibilities. Simply use the receiver to "point" to several stations in different directions from where you live. But more importantly to our purpose, directivity fashions another tool to discriminate among several signals on one frequency.

Take time to listen to the static between the various signals you hear. Notice how strong it sounds at a particular volume setting. Listen to how the static sometimes seems to rise and fall. The atmosphere and our star, the Sun, conspire to make up much of this static as well as the overall radio-frequency environment.

Assuming you've conducted this experiment at night, try the same experiment around noon. Notice how different things sound? Notice how many of the signals that you heard the night before have disappeared? Notice that the static seems louder when the sun is high in the sky? What does this show you? You have discovered that, within the mediumwave (AM) band of radio frequencies, signals tend to travel further at night with less atmospheric noise. This phenomenon is so prevalent that the Federal Communications Commission (FCC) requires certain broadcast stations to adjust their power levels in the evening so that their signals do not interfere in parts of the country they were not meant to service.

> **The Federal Communications Commission (FCC) was established in 1934 to regulate interstate and foreign communications in the public interest.**

Now this information should turn on a little light above your head. Think for a minute. The sun sets in different parts of the country at different times. This means that stations around the country will be lowering their power at different times. Try the tuning experiment one more time. This time, start your listening just before local sunset and continue until darkness falls. Concentrate on those frequencies you previously discovered to have more than one station on them. You should hear some new signals clearly before the night changes occur. Now you're discovering how human effects (the FCC directive to lower power) and atmospheric effects (the changes in propagation brought about by moving out of the sun's path) can both have a bearing on your ability to seek out signals.

In conducting this series of experiments, you have learned a teensy bit about the science of how radio frequency energy works. But you have also made some other discoveries. You have found that there is a vast world of radio monitoring beyond the half dozen stations you have listened to all your life. You have also begun to understand how to combine knowledge, skill, and patience to hear signals you might never have discovered if you hadn't taken up the practice of radio monitoring. You're on your way, my friend!

A DC to daylight frequency guide

As I stated in the introduction to this book, I fancy myself a "DC to Daylight" radio monitor. This means I enjoy listening to signals captured from all portions of the radio frequency spectrum. Not everyone who becomes involved in the radio monitoring hobby takes such a broad approach to their listening. As you begin to monitor, you may find certain portions of the spectrum appeal to you more than others. No problem, no one says you have to eat the whole apple. You can just take a bite or two and be as satisfied as I am.

A little painless physics

Before we jump headlong into a quick tour of what the radio-frequency spectrum has to offer, it might help to have an even quicker tour of how this spectrum works. Now, I am about to spout some radio theory that only scratches the surface of the full understanding of how radio signals travel from transmitter to receiver. I'll try to make this as painless as possible. Without delving too deeply into the physics of things, you need to understand that radio-frequency energy travels in waves. Think for a minute about the kind of waves you are familiar with. Think of waves crashing on the beach or the ripples formed by throwing a rock into a pond. It's fairly easy to see that these waves move through *time* over a *distance*. If you were to mark a fixed point, you could count how many waves moved past that point in a minute or even a second without too much difficulty. Radio waves do it the same way, but at the speed of light, about 186,000 miles per second, a bit faster than can be observed with the naked eye. As the frequency of waves increases, the distance between the waves decreases. So the higher the frequency, the more waves will pass a given point within a fixed period of time. The lower the frequency of the waves, the longer the distance between each wave.

Yeah, I know I just said the same thing three different ways. There is method to my madness. If you can hang onto this fact and keep it in mind as we move through the rest of this book you will already have a handle on just about the most important bit of electronic theory that a radio monitor needs to know. The unit of measurement that science gives to the movement of one wave passed a fixed point in one second is the **Hertz** (Hz). For example, a frequency of 1000 Hz means that the waves were moving at the rate of 1000 waves per second past a fixed point in space. By the way, this might be a good time to review the metric system as it applies to frequencies: 1000 Hertz (Hz) equals 1 Kilohertz (kHz); 1000 Kilohertz (kHz) equals 1 Megahertz (MHz); 1000 Megahertz (MHz) equals 1 Gigahertz (GHz).

If you are old enough to remember Howdy Doody, you may have heard the terms *kilocycles* and *megacycles* used instead of *kilohertz* and *megahertz*. Those folks who are in charge of worrying about such things decided that it might be nice to honor the German physicist Heinrich Hertz for his practical discovery of electromagnetic waves. This might be an appropriate place to try a cheap pun such as I hope your head doesn't "Hertz" from all this theory I am throwing at you, but even I wouldn't stoop to that. Don't let the theory get you down, my friend. For most of what you'll be doing, the theory you use will be used in fun ways, the rest you can always look up when you need to or want to. Remember, I said you don't need to be a rocket scientist to enjoy radio monitoring. I'm as math dumb as they come and I've held my own in this hobby for a long time.

FIGURE 2: MODULATION TYPES

No. of periods per second = Frequency (Hz)

Period

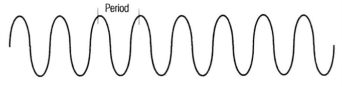

A. Unmodulated RF carrier wave

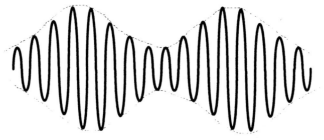

B. Amplitude modulated (AM) RF carrier wave

C. Frequency modulated (FM) RF carrier wave

Speaking of waves, there is a direct relationship between the length of a wave and its frequency. The lower the frequency the longer the wave, the higher the frequency the shorter the wave. As you move into the radio monitoring world you will hear people often refer to particular segments of the radio frequency spectrum in terms of the actual wavelength measured in meters. For example the length of a wave at 500 kHz is 6000 meters, a wave at 5000 kHz (or 5 MHz) is 60 meters, a wave at 5,000,000 kHz (or 5 GHz) is .06 meter.

You don't need to do anything with this fact for the moment, Just keep it in mind as we move on to some practical monitoring issues down the road. You will find that it is has many uses.

Let's jump sideways here for a moment and consider the human voice. You probably learned in school that we are able to speak by our vocal cords vibrating against one another and producing (you guessed it) sound waves. Sound, also known as the *audio frequency spectrum*, operates from 20–20,000 Hz. As you will see in a few moments, this overlaps with the very bottom of the radio frequency spectrum. Think of your voice as a transmitter and another person's ears as a receiver. In the case of the audio frequency spectrum, when you talk, the sound waves created by your vocal cords travel as rapid changes in air pressure that are then received by your friend's ear. The anatomy and neurology of the ear converts the audio frequency signals into something your friend's brain recognizes as your voice.

In terms of physical explanation, radio is not really that much different from sound. The main difference is that instead of sound waves, we're dealing with electromagnetic waves. If you think about what happens as you talk and people listen, what we call radio just adds a few steps in the middle. Your friendly talk show host speaks into a microphone, the microphone converts the spoken words into an electrical signal in the audio frequency spectrum and sends it to the transmitter. The transmitter converts this audio frequency signal into a radio frequency signal. The transmitting station generates an electromagnetic signal that it sends out its antenna. The signal travels through space and is picked up by the receiver's antenna. The receiver goes through a series of steps to convert the radio frequency signal back into an audio signal that you can then hear with your ears as the voice that you recognize as your friendly talk show host.

According to Article 2, Number 12, of the 1959 Geneva Radio Regulations, the radio frequency spectrum (frequencies that support moving information from place to place by way of electromagnetic waves) runs from 30 Hertz to 3000 gigahertz (GHz). 1 gigahertz is equal to 1 billion Hertz so you can see that we are talking about a lot of frequency bandwidth. Just beyond the upper limit of the radio-frequency spectrum you will find the *light spectrum*. So between voice and light is radio. The standard AM and FM broadcast bands make up a very small portion of this entire radio-frequency spectrum. You have probably been quite content with your limited use of radio. That is, up until now. If you've come this far without putting this book down, you clearly exhibit the curiosity that lies at the very heart of the radio monitoring experience.

Okay, true confession time. Probably nobody is truly a DC to Daylight radio monitor. At the lower 30 Hertz end of the band are signals that are designed to reach submarines traveling at great depths. Way up at that 3000 gigahertz end of the world you will find your microwave oven and a handful of folks in lab coats studying phenomena that they haven't even figured out yet. In either case, the equipment used to monitor at the far ends of the spectrum is usually out of the price range and the skill level of most hobbyists just starting out. But in between these two extremes is more fun than most folks can pack into a lifetime. Practically, radio monitoring for most hobbyists kicks in at about 30 kilohertz (kHz). Most hobbyist-level receiving equipment gives up the ghost at around 2000 megahertz (MHz). You might be interested to know that affordable hobby level equipment for the frequencies above 1 GHz has only come on the scene in recent years; so, in effect, the bandwidth the typical radio monitor has to monitor and enjoy has practically doubled in the last ten years. Emerging technology, largely brought forth by the efforts of those aforementioned folks in lab coats, will no doubt open new realms for the radio monitor in the near future.

> **The frequencies that make up the shortwave spectrum were originally thought to be useless for practical communications over long distances.**

Okay, so what do you already know going into this hobby? When you look at the tuning readout on a typical AM broadcast receiver, you will notice that the frequency readout usually moves from 540 kHz to 1610 or 1710 kHz. Or if you are looking at an FM broadcast band receiver, you will likely find markings indicating 88 MHz through 108 MHz. If the radio frequency spectrum runs from 30 Hz to 3000 GHz, that means that there must be quite a few things we are missing out on below, between, and above those stations that we catch on our car radios. In addition to broadcast stations, you can hear aircraft, maritime, military, government, business, amateur, weather, satellite, and industrial transmissions. You can even monitor a whole world of signals that exist on the "edge." Pirates, spies, clandestine governments, propaganda, drug trafficking, and illegal business operations can also show up on your receiver. You're about to enter a world you may not have had any notion existed before you started reading this book. Let's create in our minds an imaginary receiver that covers the entire radio frequency spectrum and spin up the dial from end to end to see what we might hear along the way. This trip through the radio frequency spectrum will give you a few notions about where you might want to first begin your monitoring practices. Remember, this is supposed to be fun folks, you're under no obligation to listen to or enjoy the same things as anyone else. Pick out what looks like fun to you and then read on to learn how to get started hearing what you want to hear.

Very low frequency (VLF): 10–30 kHz

The very low frequency area is truly the basement of radio. Very low frequency is synonymous with very long wavelengths. As I said earlier in this chapter, all you're likely to hear way down here are submarine navigation systems and a few other military applications. Both the United States and the Soviet Union use this neck of the radio frequency spectrum to play the kinds of dangerous games that have been popularized in books such as Tom Clancy's *The Hunt for Red October*. The U.S. system goes by the name *OMEGA*, and it operates on frequencies between 10 and 14 kHz. The reason these frequencies are used is because they are very effective in penetrating through sea water, allowing for continuous submerged submarine operations. This portion of the radio spectrum is more or less off limits to the typical radio monitor for two reasons. First, receiving equipment that covers this portion of the spectrum is harder to come by. Most entry-level hobby communications receivers do not tune down this far into radio's basement. Second, just about anything going on down here is seriously coded. Perhaps the easiest way to log signals down in the VLF region would be to head for your nearby Navy recruiting office and sign on for a hitch in the submarine service. By the way, the land-based transmitting stations that send signals out to the submarines have antenna installations that can be several miles in area.

At the upper end of the VLF band you can find *Standard Time and Frequency Stations*. These exist for the purpose of performing various calibration procedures for their users, usually the military.

Low frequency (LF): 30–300 kHz

At least a portion of the low frequency band is easily reached by the hobbyist. This is because many modern hobby-level communications receivers begin their frequency coverage at 150 kHz. The lower end of the LF bands is populated by more Standard Time and Frequency Stations as well as by a host of military operations. At 100 kHz you will find the signal for the *LORAN-C Navigation System*. If you own a boat (also known as a hole in the water that you keep throwing money into) or if you have a neighbor who is a boat owner (recognizable by the threadbare clothing brought about by throwing money into that hole in the water), you have probably heard of and possibly depended on the LORAN system. Above 100 kHz you will find more military stations, including systems that send weather charts by fax, very similar to the fax machines you might have in your office only utilizing radio signals instead of signals that come over the telephone.

As we round the corner into the 150 kHz range where hobby receivers begin to earn their keep, we run into a lady named *GWEN*. From 150 through 175 kHz the *Ground Wave Emergency Network* (GWEN) is a nationwide system established by the United States Air Force to provide for a "survivable" communications system. GWEN is there to keep things running if "The Big One" ever drops. Again, as with so many military operations, you can listen but you won't hear very much. The signals are encrypted.

From 160 through 190 kHz is an interesting place. This is referred to as the 1750 meter band. Remember what we said about meter measurements before? 170 kHz works out to a wavelength of 1750 meters. This region is also known as the "Land of the Lowfers." *Lowfers* are radio hobbyists who set up low-frequency stations to transmit signals for the fun and enjoyment of other Lowfer listeners. For some hobbyists this is a great way to get involved in putting out their own signals without needing to obtain a license to operate, as is required on the amateur radio frequencies further up the bands.

At around 200 kHz, things pick up a bit for the radio hobbyist. From 200 kHz through and beyond the 300 kHz that marks the top of the LF band, you will find a world populated with hundreds of beacon stations. Non-directional beacons are used throughout much of the world as homing signals for aircraft. They are very easy to find and can be great fun to log. The beacon station signals consist

of between one and three letters sent repeatedly in International
Morse Code. The code speed is very slow so it is easy enough to
write down the dots and dashes and look the letters up on a code
chart. When conditions are right (the best time being at night in the
winter) these signals can travel over great distances, making beacon
logging a very challenging radio monitor activity. Chasing beacons
is one aspect of the area of the radio monitoring hobby known as
Utility listening. This is where the monitor seeks to listen to signals
that do not fall in the realm of broadcasting for public consumption.
As we move on in our frequency study you will discover that many
exciting things to hear fall into the Utility listening category.

If you live on the eastern coast of the United States, on a cold
winter night it is possible to hear some European-based broadcast
stations from 155 through 281 kHz. These stations are similar to
those we hear every day in the standard AM broadcast band, except
that you are likely to hear a language other than English. Low Band
broadcast stations are rare finds and great additions to your log.

Medium frequency (MF): 300–3000 kHz

With medium-frequency signals we get a bit closer to radio as
you probably envisioned it before you joined folks like me in the
radio monitoring world. Still, there are a few matters to get excited
about before we run up against the standard AM broadcast band.
You will still find those non-directional beacon stations that we just
talked about happily beeping away in the International Morse Code
all the way up through 415 kHz. There are a few other signals
interspersed within these beacon stations. From 415 - 515 kHz is
maritime communications. This chunk of the band is in a state of
flux, as almost all communications here were previously in
International Morse Code, including the maritime distress frequency,
512 kHz. The trend is to move away from Morse code as a primary
means of maritime communications, so look for changes in this area.

From 515 kHz through 540 kHz brings us another group of
beacon stations similar to those we mentioned earlier. Some of these
are even easier to hear than their lower-frequency counterparts.

From 525 kHz through 535 kHz you will find *Traveler's
Information Service* (TIS) stations. You have probably run across

these stations from time to time if you travel by car around the country. Often they are set up in relation to major highways with potential traffic problems or construction sites. They can also be found at the entrances to parks, beaches, museums, or amusement areas. These are relatively short-range transmitters that send out a signal by way of a repeated taped message, letting folks in on what's going on around them. When conditions are right, usually in the evening, these signals can sometimes travel fairly long distances. Next time you are in your car at night, tune down to the lower end of the band and see if you can hear anything. You may be surprised.

In December, 1995, WJDM, Elizabeth, NJ, became the first broadcaster in this new segment on 1660 kHz.

From 530 kHz - 1610 kHz we find familiar ground. This is the good old *Standard AM Broadcast Band*. Up until you became interested in radio monitoring you probably took the AM band for granted just like so many other folks. Once you programmed a few stations into the buttons on your car radio, you probably never gave signals here another thought. The AM broadcast band is a signal-rich radio-frequency environment with enough activity to generate its own substrata in the radio monitoring world. There are many folks who find happiness in logging signals on this band and never even feel the desire to head anywhere else. AM broadcast band monitoring has so much potential for the beginning hobbyist that I have devoted an entire segment of the book to it beginning with Chapter 5. We will talk more about this aspect of the hobby later, but I do want to whet your appetite. Did you know it's possible to log in excess of 1000 stations on this band? Did you know you could log foreign countries on the band? Stick around my friend, we have much to discuss.

A recent development in the Medium-Frequency band is the Federal Communications Commission's (FCC) plan to expand the AM Broadcast Band. Stations are currently applying for new license allocations that will expand the existing AM broadcast all the way out to 1700 kHz. Several stations have already begun broadcasting in this new band segment. This will further increase interest in this band to many radio monitoring enthusiasts.

While we are waiting for this new development to bear fruit, the current upper end of the Medium Frequency Band mirrors its lower end. At 1610 kHz you will again find Travelers Information Service stations similar to those around 530 kHz. From 1610 through 1700 kHz you will hear a few more beacon stations, mostly from South America, which make them nice catches for the log. In this range you may also hear a few **Pirate Broadcasters**. We will discuss pirates in more detail in Chapter 28, the shortwave section.

From 1700 kHz through 1800 kHz is a segment of frequencies that were at one time fairly well populated by cordless telephones. These frequencies have fallen out of use somewhat, and current government regulations discourage monitoring of cordless-telephone frequencies anyway.

From 1800 through 2000 kHz you enter the world of **Amateur Radio** for the first time and this is an aspect of the radio hobby that you may find interesting. The hobby consists of people who have been licensed by the FCC (or the equivalent government agency in other countries) to set up their own stations for the purpose of communicating with one another. Amateur radio is a lot of fun to do (and we will talk about getting in on the fun later) but for now I want to let you know that these can also be fun signals to listen in on as a radio monitor. These are also the frequencies at which you will first begin to encounter signals that sound like a duck quacking and trying to talk. Those are **Single Sideband** (SSB) signals and we will be talking about how to turn them into voices you can understand later on in the book in the chapters devoted to shortwave listening beginning with Chapter 20. SSB is a very popular mode of communication in the non-broadcast radio world. Some overseas broadcast stations have also experimented with using SSB. Don't worry too much about this. Most modern general-coverage communications receivers that tune through ranges where SSB is in use have features that allow these signals to be heard properly.

From 2000 kHz through 3000 kHz you will mainly find **Maritime Communications**. If you become interested in listening in on ship communications you will always keep one ear on 2182 kHz. This frequency is universally recognized for emergency distress and safety communications. From 2300 kHz through 2495 kHz is also recognized as the first of the **World Broadcasting Bands,** known as the 120 Meter Band. This is also the first of what is known as the **Tropical** bands. In the part of the world we live in, we are used to the standard AM and FM radio bands as the primary sources of radio. In many other parts of the world, mainly those grouped toward the equatorial regions of our planet, the tropical bands are used for wide-ranging domestic broadcasting of interest to the various country's local populations (and of interest to many dedicated radio monitors). Some of the most challenging and rewarding long-distance listening (DXing) can be done on the tropical bands. Also, at 2500 kHz you may be able to hear the first of several easy-to-log **Standard Time and Frequency Stations**. Here you may hear the signal of WWV, Fort Collins, Colorado or WWVH, Kauai, Hawaii. WWV and WWVH also broadcast in the HF spectrum at 5, 10, 15, and 20 MHz. Once you find a regular time signal you will become obsessed with having the accurate time. You will also be totally frustrated with the rest of the non-radio monitoring world, who just can't seem to be on time for anything.

> **The National Bureau of Standards "atomic clocks" heard on WWV and WWVH are so accurate that an occasional "leap second" must be added to adjust for the slowing of the earth's rotation.**

Since you have the most accurate time, why not listen in on the latest weather report. On 2670 kHz the United States Coast Guard broadcasts regular weather bulletins of interest to ships and coastal areas. This signal is broadcast in SSB.

From 2850 kHz through and above 3000 kHz you will hear **Aircraft Communications**. Aero monitoring is another of those aspects of radio monitoring that has developed its own substrata of enthusiasts. Aircraft listening is one of the few areas of the hobby that encourages monitoring in at least three different major frequency regions. Aircraft listeners monitor the above mentioned frequencies in the Medium-Frequency region. They also have plenty to hear in the High-Frequency region and in the Very-High-Frequency regions we will be discussing later in this chapter. Radio monitors dedicated to aircraft listening come as close to being true "DC to Daylight" listeners as any group in our hobby. And they never have to leave the ground. Imagine that!

High frequency (HF): 3 MHz – 30 MHz

Now we move into that world between the more familiar AM and FM standard broadcast bands. The **High-Frequency** portion of the radio spectrum is the segment that is commonly known as the **shortwave** band. The roots of this name go back to the early days of radio when most operations were in the lower-frequency, longer-wavelength bands. Migration to the shorter wavelength bands brought reliable world-wide radio communications. Hearing these frequencies requires a **general coverage receiver**, but once you start listening, you'll never want to stop.

Radio Moscow became the Voice of Russia after the USSR collapsed

Shortwave monitoring is another of the major topic areas we will pursue later on in the book, beginning with Chapter 20, so let me just quickly tell you what you will hear on this band. YOU WILL HEAR EVERYTHING!!!!! There are international and worldwide domestic radio broadcasts; utility stations conducting the world's business; as well as military, government, police, and emergency services. You will hear ships, aircraft, and even spacecraft. You will hear licensed amateur radio operators from around the world as well as non-licensed clandestine and pirate radio stations. You will hear transmissions in the normal voice mode similar to what you hear when you listen to your AM radio in your car.

You will also hear many other modes of broadcast including **Single Sideband** (SSB), **Continuous Wave** (CW) more commonly known as International Morse Code, **Radio Teletype** (RTTY), **Facsimile** (FAX), and even more exotic ways of communicating by way of radio around the world. You will find ten segments of this band dedicated to shortwave broadcast stations, eight segments devoted to amateur radio, plus the traditional **Citizen's Band** (CB) and everything in between occupied by every kind of signal imaginable in the classification of utility stations.

Many hobbyists choose to concentrate their monitoring practices on just the shortwave bands; some find the signal environment so rich that they further specialize into monitoring just broadcast or utility or other signals. As a beginner you will probably want a taste of everything at the table. Down the road you can decide what your favorite dishes will be. The scope of what can be heard in this segment could cover several books. It would be easy for a beginner to become overwhelmed, but the book you hold in your hands will help you sort your way through to fun monitoring right now.

The crossover point

From the top end of the HF band through the bottom end of the VHF band, radio begins to make a critical shift at around 30 MHz. Some of the basic principles of propagation, transmission, and reception begin to require different thought. For example, often atmospheric conditions that affect the lower frequencies negatively can bring about enhanced conditions in the VHF and above ranges. The other major shift is many of the signals are no longer transmitted by way of **Amplitude Modulation** (AM). Simply put, AM means the information that we get from a radio signal is carried on the signal by varying the height or amplitude of the wave. FM stands for **Frequency Modulation**. In FM the information that we get from the radio signal is carried on the signal by slightly varying the frequency of the waves. As a beginning listener, even this much information about the subject is little more than a curiosity. Modern receiving equipment makes switching between modes of transmission as easy as pushing a button.

While we're on the subject, don't fall into the trap of getting bogged down in the minute details of radio technology. At this stage of the game you can get along fine without them. As you mature in the hobby you will find that a passing knowledge and a book or two will keep your head above water. If you are excited about the electronics and theory of radio, by all means pursue this interest.

I'm simply letting you know that, as a beginner, you'll probably discover much of what you need to know by simply spending your time actually listening to your radios instead of worrying a great deal about how they work. End of sermon.

The other major shift that occurs is that most VHF and higher radio monitoring takes place in the frequency ranges covered by scanning receivers. Hence, the need to buy another receiver. Only recently have receivers that cover both the frequencies below and above 30 MHz become available. These receivers are still a somewhat expensive investment for a beginner. In most cases the two receivers needed to span the gap are still less expensive than the single one that covers both HF and VHF/ UHF. Some of these "extended coverage" receivers also contain design compromises that still make them favor one side or the other of 30 MHz.

So you might see 30 MHz as the radio monitoring hobby's great divide. Most folks begin their listening above or below this point. Some folks never cross over. But "DC to Daylight" listeners span this gap without giving it a great deal of thought. As you begin your monitoring experience, you will discover which path is right for you.

Very high frequency (VHF): 30–300 MHz

30–50 MHz

This VHF-Lo band has a wide array of signals from the government and business worlds. The tendency in recent years has been for operations to move into the upper end of VHF and into the UHF regions. Still, no frequency goes unused in a world that runs on radio. Down here in the VHF basement are law enforcement, the Red Cross, highway maintenance, forestry, utility, and petroleum businesses. You will even find children's low-power "walkie-talkies" and some cordless phones. Let me state again that it is illegal to monitor cordless phone conversations in some states.

50–54 MHz

This is the **6 Meter Amateur Radio Band**. Hams use it for traditional communications but they also use a small segment for remote control and operation of model airplanes, boats, and cars.

Television: 54–72, 76–88 MHz

Television channels 2-6, occupy these VHF-Lo band segments. The audio and video portions of the radio signal that we call television are each transmitted on a separate frequency. As you listen through these frequencies you will hear the "voices" of television interspersed on frequencies between other frequencies where you will hear buzzing noises. That buzzing sound is the video portion of the TV programming. There are folks who monitor TV and attempt to receive TV stations from long distances. Just another fascinating facet of the radio-monitoring world.

72–76 MHz

This is an interesting corner of the spectrum. In addition to paging transmitters and some industrial communications, these frequencies are popular with surveillance and eavesdropping equipment users, such as law enforcement agencies and private investigators.

88–108 MHz

From 88 to 108 MHz we return to a more familiar world, the **Standard FM Broadcast Band**, or so you may think. Many FM stations also broadcast **Subsidiary Carrier Service** (SCS) signals underneath the programming you normally hear. This SCS music, programming, and data can, by law, only be monitored by people who have the transmitting station's permission to do so. However, equipment to receive these signals is within the budget and expertise of even beginning radio monitors.

108–137 MHz

From 108–137 MHz you will find **civilian aircraft** operations. These signals are primarily AM instead of FM for reasons of safety.

FM receivers exhibit a property called "capture effect" where the strongest of two or more signals on a frequency is dominate in that receiver. While this is highly desirable when traveling in your car listening to your favorite music program, it can spell disaster to aircraft. By using the AM mode, weaker signals can still be heard beneath stronger signals. This means that signals from a plane in trouble have a much better chance of being heard by others. As I stated earlier, monitoring aircraft can be really exciting.

137-138 MHz

Monitoring 137-138 MHz allows you to become a "Space Cadet." These frequencies are used by many weather satellites. While you won't be able to directly interpret the sounds that you hear, hobbyists who become interested in satellite transmission can purchase equipment that allows the signals received to be translated into weather maps. These **WEFAX** pictures can be translated by using special demodulator units that connect to any receiver covering these frequencies.

138–144 MHz

This band is allocated to the government, and is used primarily by **military bases** for security and field operations. There may be some experimental research as well as Civil Air Patrol in this band.

144–148 MHz

The 144–148 MHz portion of the VHF spectrum is known as the **2 Meter Amateur Radio Band**. This is arguably the most popular place in amateur radio. On these frequencies, amateurs establish high–powered "repeater" stations that allow them to communicate with each other over great distances using relatively low-powered and inexpensive handheld equipment. Hams use this frequency range to establish emergency service communications to assist public safety professionals in times of local emergency and disaster such as floods, hurricanes, or hazardous material spills. Monitoring these frequencies can bring some very exciting signals into your log.

144-148 MHz (2-meters) is the most popular amateur band
(Photo courtesy of Icom America, Inc.)

148–300 MHz

From 148 to 150.8 MHz can disclose more government/military signals including those of the **Civil Air Patrol** (CAP) and the **Military Affiliated Radio Service** (MARS).

150.8–174 MHz is the most signal-rich area in VHF radio monitoring. You will hear all the business, government, public safety, and law enforcement signals traditionally associated with listening to what people have come to call (wrongly, I might add) a **police scanner**. In this range you will also find boats, trucks, railroads, hot air balloons, and press services. Just about anything on the move that requires radio to get the task accomplished can be found somewhere in this section of frequencies.

174–216 MHz are VHF-Hi TV broadcast channels 7–13, not significantly different from the description for channels 2–6.

The frequencies from 216 through 220 MHz are used for inland waterway communication while the 220–225 MHz portion is utilized by the land mobile services on the lower end and shared with amateur radio operators on the upper end. Amateur practices in this band are similar to the earlier mentioned 2 meter band.

The frequencies from 225–300 MHz round out the VHF band with military aviation frequencies. Again AM mode is used on these frequencies for safety reasons. Military satellites also are in this band with various modes of modulation.

Ultra high frequency (UHF): 300–3000 MHz

The military aircraft band beginning at 225 MHz extends through 400 MHz to open the door to UHF. 400–420 is a federal government band, though 400–406 MHz is likely to be shared and populated with various kinds of telemetry signals. This is a popular area for weather balloon experiments. From 406 through 420 MHz is where you can hear the FBI, DEA, and even **Air Force One** if the President of the United States is visiting near your site.

420–450 MHz

From 420 through 450 MHz is yet another popular Amateur Radio band. In addition to traditional communication you will discover hams on this band experimenting with television, satellite communications, and even bouncing signals off the moon.

450–470 MHz

This band takes on many of the same roles and characteristics as the 150.8–174 MHz band. Many services on the VHF band have moved up to these frequencies as operating costs decreased. This just means more signals to add to the log.

470–806 MHz

This is the UHF television band. Some areas that don't have TV on channels 14–20, are permitted to use 470–512 MHz for other purposes including public safety.

806–824 MHz

This band is allocated to mobiles in the Business, Public Safety, and Land Mobile Satellite Radio Services, some of which is trunked.

824–851 MHz

This is the *mobile* **cellular telephone band**. Monitoring these signals is illegal. We will discuss this and other prohibitions in Chapter 57, entitled *Listening, the Law, and Common Sense*. Don't worry, there are only a few places you aren't supposed to listen.

> **Many people who use cellular phones, cordless phones, and "baby" monitors forget that these signals can be intercepted with common receiving equipment.**

851–869 MHz

This is the repeater or base output band for the 806–824 Business, Public Safety, and Land Mobile Satellite Radio Services.

869–894 MHz

This is the *base* **cellular telephone band**. Monitoring these signals is illegal. We will discuss this and other prohibitions in Chapter 57, entitled *Listening, the Law, and Common Sense*.

894–1300 MHz

894–902 MHz	Mobile band for the Private Land Mobile Service.
902–928 MHz	The new cordless telephone band.
928–932 MHz	Common carrier pagers operate in this band.
932–935 MHz	Federal government, but possibly shared with other services.
935–941 MHz	Repeater output or base band for the 894–902 MHz Private Land Mobile Service.
941–944 MHz	Another government band, possibly shared with other services.
944–952 MHz	An interesting band, albeit *illegal to monitor*, allocated to the Broadcast Radio Service. It's used for signal relay, typically studio-transmitter links (STL).
952–960 MHz	Here is an uninteresting mix of the Private Microwave and Private Fixed Services, probably illegal to monitor anyway.
960–1215 MHz	Allocated to the aviation services, typically for TACAN and DME (distance measuring equipment). Signals are probably all non-voice and encrypted or uninteresting.
1215–1240 MHz	This is sort of an extension of the 960-1215 MHz band, used for radiolocation and radionavigation equipment.
1240–1300 MHz	Here we have another Amateur Radio band, affectionately called the 23-cm band. It's quite active in some parts of the world.
1300 MHz–3 GHz	Frequencies above 1300 MHz represent the new "high frontier" for the radio hobbyist. Some of the latest receivers are designed to catch this new world, diversely allocated to radio astronomy, earth exploration, data telemetry, radionavigation, television broadcast relay, satellites, world-wide maritime communications (INMARSAT), space research, meteorology, and much more.

Hams are active in the 1200 MHz band
(Photo courtesy of Icom America, Inc.)

Super high frequency (SHF) :   3 GHz–30 GHz

Extremely high frequency (EHF): 30 GHz–300 GHz

Practical radio in the SHF and EHF regions still sits on the edge of reality for all but a few dedicated radio-frequency engineers and a handful of visionaries. Much of the activity up here is experimental, amateur, or government at this time. Still, the 3 GHz–30 GHz segment is sprinkled with applications not unlike what we see at lower, more traditional frequencies. Just remember that people once scoffed at radio being "useful" above what we currently know as the Standard AM Broadcast band. My first VHF/UHF scanning receiver in the mid-1970s covered frequencies up to 450 MHz because "nothing was above that point." In anticipation of the future, this segment of the spectrum is already divvied up including broadcast segments at 40.5–42.5 GHz and 84–86 GHz. Get ready for the future! When the signals move up the dial, you can be among the first to hear them, just like those dedicated few who heard Marconi's "S" in 1901 or those who strained to pick KDKA out of the static in 1920. It will always be an exciting world for the radio monitor.

> See Appendix 4, "*Radio Spectrum Allocations*," for a detailed guide to the monitorable RF spectrum. It's drier and more specific than this chapter, but should prove to be a handy reference.

5

Mediumwave monitoring — a great place to start

At this point it might be helpful to run down a bit of my pedigree. I've already told you I have been in the radio monitoring hobby for more than twenty-five years. In addition to "playing the game," I have devoted a portion of my time to helping beginners get started in the hobby. I have done this mainly through writing the "Beginner's Corner" column for *Monitoring Times* magazine since 1988. In that time I have also written articles for other hobby-related magazines and journals, always with the beginner's needs in mind. I've even conducted beginner's seminars and workshops at radio conventions, ham radio meetings, and other venues. During my tenure as a beginner's writer I have corresponded via postal mail and E-mail with many beginners about their individual hobby concerns.

So much for my background: what's the point? Well, when I first got the notion to write a book for beginning radio monitors, I took a look at the existing literature on the subject and compared what I was reading to what I was hearing from the beginners themselves. The result of this information is this section on mediumwave listening. All too many beginners' books up to this time have placed MW at a fairly low priority, often making it a secondary concern to shortwave listening. I decided things should be different.

Mediumwave listening is a great way to enter the radio monitoring hobby for two simple reasons.

1) Most folks already have a Standard AM broadcast Band receiver in their home or car. Therefore, start-up expenses are essentially nothing or very small.
2) No additional special equipment is needed from the start. You can begin the radio monitoring hobby RIGHT NOW!!!

Of course this does not mean you will necessarily choose to stay with mediumwave monitoring, although some folks do make MW their primary hobby activity. Mediumwave listening gets you developing your radio monitoring "chops" while you decide what other aspects of the hobby excite you. Also, since shortwave and VHF/UHF monitoring will require that you purchase a specialized receiver, listening to the AM broadcast band gives you something to sink your teeth into while you're developing the resources to move on to other areas of monitoring interest.

The Standard AM Broadcast band gets taken for granted. Maybe this is because it has always been there, somewhere, in our lives. If you were to take a poll in your neighborhood, chances are everyone over the age of 5 years old will be able to tell you the call letters of at least one AM station. Why did I pick that age? Because that's when kids begin to sit glued to the radio on a snowy winter morning waiting to hear if school is closed. You folks who live in warmer climates probably have similar behavior relating to other weather conditions. But the point is that AM radio has always "just been there" for most of us. But things are different once someone decides to become a radio monitor. Then, even the common and familiar Standard AM Broadcast band takes on a different appearance. Hunting and poking around between those few familiar stations reveals that this band is full of monitoring fun.

6

What can you hear on mediumwave?

Go to your nearest AM radio and turn it on. Spin slowly up and down the dial. You will probably hear about half a dozen kinds of music, another half dozen talk shows, a sports show or two, a religious broadcaster or two, and maybe a news station. You may even hear one or two stations broadcasting in Spanish or some language other than English. Fairly familiar stuff, huh? Okay, that's on the receiving end of things. Try to imagine things from the point of view of the broadcaster. Those folks on the other end of the microphone are trying to communicate with as many people as they possibly can over an area as large as their signal allows. The word *broadcasting* actually predates radio. Broadcasting originally referred to a method of sowing seeds to cover the largest area of land possible with the least amount of work. Now the term refers to cultivating a particular reaction from a group of listeners, rather than cultivating corn. In the case of most domestic radio broadcasters, the reaction they are seeking is that you will either buy their sponsor's products or buy into their way of thinking about some subject or topic. "Commercial" broadcasting is often the key to success in the mediumwave listening world. Many stations these days reduce overall costs by using programming that they receive "canned" via a tape recording, satellite, or other feed. They simply insert their local information and commercials in between portions of these program feeds. It is possible to tune across the AM band and hear the same talk show being broadcast on a number of stations. You need to listen through to either the top of the hour to hear the station's call sign, or you can pay attention to the commercial breaks to get an idea of where the signal is coming from. So the radio monitor isn't just interested in the program content. The radio monitor wants to know where each particular signal originated. Hearing a talk show is entertaining. Hearing that same talk show simultaneously broadcast from stations in several different states is radio monitoring.

It's possible for a person monitoring the AM broadcast band to hear stations in all 50 United States, all 10 Canadian Provinces, all 10 Mexican Estados, a number of Caribbean and South American countries, as well as a few transatlantic or transpacific countries depending upon which coast you're closer to. Loggings of over 6000 miles have been verified. All you need to do to hear some of these wonders is to listen to your AM receiver a little bit differently than you might have in the past. You need to listen as a radio monitor.

Let's start out with a look at what can be heard by monitoring the domestic AM broadcast stations. There are three major classifications of stations licensed to operate in the United States.

Clear channel stations

The first type are known as clear channel stations. They operate with a power level of up to 50,000 watts, the highest power currently allowed by law on this band. Most of these stations operate at this power level 24 hours a day and their signals can be heard over great distances. There are currently 60 Clear Channel frequencies authorized. They are spaced every 10 kHz and can be found on 540, 640–780, 800–900, 990–1140, 1160–1220, and 1500–1580 kHz. The name Clear Channel comes from a time when stations assigned to these frequencies had essentially exclusive use of the slot.

50 kw / 1210 kHz "Clear Channel" station WPHT

Since 1980, the Federal Communications Commission (FCC) has allowed other lower-power stations to also operate on the same frequencies as the traditional "Clears." As a monitor, this is essentially a case of looking at the glass as half empty or half full. It is harder to log some of those traditionally easy long-distance signals, but now there are more potential possibilities to log new stations. As you become more adept at monitoring, you will keep an ear out for stations that normally run 24 hours, going off the air for maintenance. This usually occurs late Sunday evenings or at other times of low listenership. When one signal goes off the air, you will usually have the opportunity to log one or two more that you didn't hear in the past. As a beginner, it is a good idea to begin your listening on the clear channels. You will be able to log many states during the first few weeks of listening by concentrating on these frequencies. Also, wrestling with the interference brought about by the lower-power station will help you to train your ears for digging out weaker signals on some of the channels we will be talking about in a few minutes. By the way, our neighbors in Canada, Mexico, Cuba, and the Bahamas also have identified clear channel frequencies that make logging these other countries a breeze for even a beginner.

Remember the experiment we conducted on the directionality of radio signals back toward the beginning of the book? As you begin to listen in on the clear channels, you can begin to apply this practice of turning your receiver from side to side to produce "nulls" that will help you hear more than one signal on each frequency. If you discover you are having so much fun hearing new things, don't be afraid to just put this book down and monitor for a while. You really can't do anything wrong. The world will not come to an end if you hear one station and not another. There will always be another chance to log something. Go ahead and play, it's perfectly okay. I'll be here when you get back.

Regional channels

The second type of domestic AM station is known as a regional channel, occupied by stations with a maximum power of 5000 watts in the daytime, substantially less than a Clear Channel station. At night they are even more severely inhibited, with limited power authorization of between 500 and 1000 watts with a directional antenna to reduce the chance of interference with other stations. Instead of being established to cover a wide area, these stations usually service, for example, an individual town or city and its immediate surroundings. There are currently 41 regional frequencies authorized. They are spaced every 10 kHz and can be found on 550–630, 790, 910–980, 1150, 1250–1330, 1350–1390, 1410–1440, 1460–1480, 1590, and 1600 kHz. Once you have cut your teeth on the clear channels you will be ready to accept the challenge of the regional broadcasters. These frequencies will present you with many exciting signals from places you never expected your humble AM receiver to hear. The regionals will make your log book bulge.

Graveyard channels

Think you've heard everything? We're just getting warmed up, my friend. Now its time to take a walk through the "Graveyard." These frequencies may have come to be known by this term because they were supposed to be "dead" at night. To the radio monitor they are far from dead. In some ways they are the most lively frequencies you can find in the entire radio hobby. The graveyard frequencies are just six in number: 1230, 1240, 1340, 1400, 1450, and 1490 kHz. Some radio monitors never tune anywhere else. Why all the excitement? These six frequencies are known as local channels. Stations on these frequencies are limited to a maximum power level of 1000 watts. You probably have more light bulbs going in your house right now than that. These stations are designed for extremely local service. As such, each of these frequencies has more than 120 stations assigned to it. The graveyard channels may be designed for local communication but a patient radio monitor can dig out signals from great distances with a little tenacity and technique. Serious practitioners of the graveyard monitoring art can have well over twenty stations logged on each of these frequencies alone. It represents one of the biggest AM listening challenges.

Subclassifications

Altogether there are 20 separate sub-classifications referring to the three major areas mentioned above. These refer to a station's combination of power level, hours of operation, and antenna pattern. As you move on in your listening, knowing these details can help you further your quest for elusive signals. Later on, in Chapter 16, we will talk about where this information can be acquired. One of the things you will find important in pursuing AM monitoring is the need for timely and accurate station data. This is because stations have lately been changing their call signs, program formats, and operating practices about as often as most folks change their socks.

Signal spacing

Stations in the United States, Canada, Mexico, and South America all use a standard signal spacing of 10 kHz. This makes things nice and tidy in this neck of the woods. However, other parts of the world use a signal spacing of 9 kHz. So what does this mean to us? Well as you begin to tune around listening for the "normal" signals you expect to hear, peeking in and out from between those 10 kHz spaced stations just might pop a foreign nation or two. You see, there is all this stuff out there that you probably didn't even know existed and it's all waiting for you to listen in.

7

Mediumwave cost and budget issues

Throughout your participation in the radio monitoring hobby you will always need to divide your hobby budget into the areas of *Receivers, Antennas, Accessories, Information Management,* and *Information Acquisition.* You don't need to spend money on everything at once. But you will need to keep these budget categories in consideration.

Exploring the AM Broadcast band is a great way to get started in the radio-monitoring hobby with little or no financial burden. If your curiosity got the better of you and you tried some of the things we discussed earlier, you may have already discovered that you have an AM receiver that will get you started. Literally, at the initial stages of involvement, any modern AM receiver and a notebook to log what you have heard will just about cover the start-up costs of becoming a radio monitor. It is possible that you can remain happy with "existing" equipment for quite some time. As you gain experience with the club and publication resources we will talk about a little later in Chapter 14, you will become more familiar with the receivers that have proved themselves to be especially adept for AM listening. Also, as you go digging for the more difficult signals, you will find the need to resort to one of several designs of antenna systems. Again, the good news is that antennas for serious AM monitoring can be inexpensively produced for less than the cost of a trip to your local fast food restaurant. No accessories are needed at the beginning stage of AM broadcast band monitoring; however, if you look around your house, you may find you are also already in possession of the foremost listening accessory, a tape recorder. Sometimes when you are trying to decipher several signals on a single frequency, reviewing them on tape makes sorting things out much more easy. Information management consists of your log book. This can be a simple as a loose leaf notebook. I know of one long-time listener who has used stenographers notebooks as logs throughout a 30-plus-year history in the hobby. Again, the expense in this area can be minimum. Information acquisition will usually involve subscribing to a mediumwave-listening oriented club publication. You are still in the price range of a trip to your local drive-through burger joint. But the information received from these club publications will provide lasting benefits, unlike the empty calories at the fast-food emporium.

As we look at the finances attached to the hobby, you will discover that folks involved in this hobby have spent from nothing all the way through to equipment that cost more than your family car. I hope to be able to convince you that all the big-buck equipment in the world is no substitute for sitting at the dials and diligently listening. In the radio monitoring hobby you will quickly discover that money literally isn't everything. You will learn that there are challenges that do not respond to having money thrown at them. The rewards in this hobby come to those who work toward developing their listening skills, not to those who are intent on lightening their wallets at every opportunity.

Receiver choices for mediumwave listening

As already stated, you will probably be able to wring quite a bit of monitoring out of whatever receiving equipment you have around the house. Go ahead and enjoy the hobby with what you've got. But you are also, at this point, probably curious enough about this aspect of the monitoring hobby to wonder about the better receivers.

When shopping for a monitor receiver, two primary issues have to be considered: first, **sensitivity**, the ability of a receiver to detect and amplify weak signals. Second, **selectivity**, the ability of a receiver to accept a desired signal to the exclusion of adjacent, undesirable, and potentially interfering signals. Back when receivers used vacuum tubes, **stability** was also a major consideration. This was due to the nature of tube-based equipment to have frequency "drift." Modern solid-state equipment has virtually eliminated this concern. Still, keep stability in mind for future reference because, as we shall soon see, some of that old tube gear is great for monitoring.

Automobile radios

Okay, you've been monitoring and having a ball for a couple of weeks now with your family's "go to the beach" portable receiver. Still, you yearn for something with better performance possibilities. But you don't want to lay out a lot of cash. My suggestion? Put down the family beach radio and head for the family car. Most modern car radios are marvels of design. Think for a minute: a good car radio has to be designed to allow you to carry a station's signal on the move. For this reason, high-quality automobile radio receivers are designed to have high performance in the areas of sensitivity and selectivity. So before you head out to buy something new, give the car radio a spin. You may be surprised at how well this works. One well-known AM monitor used to log transatlantic signals while commuting with only the radio in his Ford Pinto.

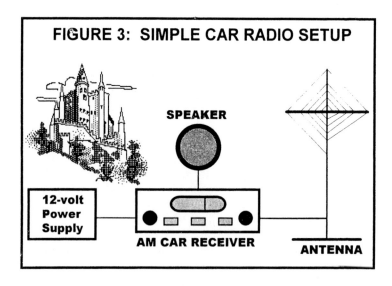

FIGURE 3: SIMPLE CAR RADIO SETUP

SPEAKER

12-volt
Power
Supply

AM CAR RECEIVER

ANTENNA

Automobile radios also possess a quality that all but a few traditional portable AM receivers lack. That is, that they are designed to be used with external antennas. Remember, I mentioned earlier on that you could do some great things with inexpensive antennas in AM monitoring. Some folks in this area of the hobby do a lot of serious monitoring on car radios they have purchased from automobile junk yards. Such receivers can usually be had in working order for between $5 and $25. Based on the exploits of the gentleman mentioned earlier, you might want to keep an eye out for receivers taken from Ford Pintos. Any high-quality automobile receiver should work fine for this project. They are easy to power by either batteries or by a simple power converter that can be purchased at any electronics store. All you have to do is wire up a speaker and one of the antennas we will discuss later in this section and you will have a system that may not be pretty, but you will hear things you may have missed with anything else you have been using. If you're handy with wood you might make a nice case for everything and turn the whole project into a great conversation piece.

Portable receivers

The next step in finding receivers that will do an even better job at AM monitoring is to seek out equipment that was designed with long-distance AM listening in mind. One such portable receiver currently in production is the General Electric Superadio. This is a high performance, large-case, portable AM/FM receiver that is highly regarded in Broadcast Band monitoring circles. In addition to its great circuitry, its two speakers and bass/treble controls produce excellent sound quality as well. The Superadios are among the few portable receivers in this class that come with connections for external antennas. Just the thing for the hobbyist. The current version — the Superadio III — retails for about $60, but can be found at discount stores for as little as $45. This receiver is a good example of the features you'll want in a serious receiver designed for the radio monitor, regardless of the type of listening you do: (1) Connections to allow the use an external antenna; (2) External audio jacks for headphones and tape recorders.

Two other portable receivers were marketed up until the mid-1980's that were practically designed for AM radio monitoring. The Radio Shack "Realistic" Long Distance AM Radio (sold under model numbers 12-655 and 12-656 and known to hobbyists simply as "The TRF") might have been the hottest AM portables ever produced. (*See preceding page.*) The other receiver is the Sony model ICF S5W. This receiver had a rather limited run but was also very high performance. Keep an eye out at flea markets and swap meets for these two rigs. You may find real diamonds in the rough.

General coverage receivers

The next step to improved performance is a general coverage receiver, 150 kHz - 30 MHz, probably what you think of when you hear the term "shortwave radio." High-performance portable or desktop receivers like these are discussed in detail in Chapter 23. If you purchase such a receiver with AM listening in mind, make sure it does in fact cover the standard AM broadcast frequencies of 530 kHz through 1700 kHz. In the past, some general coverage receivers omitted AM band coverage. Read up on all receiver specifications to assure that you don't leave out the bandwidth you are looking for.

Vacuum tube receivers

As you begin to gather information from other radio hobbyists as we will discuss a bit later, you'll find another group of receivers that is popular with AM monitors. These are what we affectionately refer to as **Hollow State** radios. The late 1950's through the 1960's produced some of the most incredibly well-engineered receiving equipment ever designed. However, the technology of the time was that of vacuum tubes. Some of these receivers can still run rings around all but the most expensive and exotic modern rigs.

Two receivers of this era are highly coveted by AM monitors. The Hammarlund HQ-180A is still the receiver of choice among some of the most dedicated AM listeners. These rigs probably logged more AM stations than any other receiver ever built. Another old timer is the military surplus receiver, R-390A/URR, designed by the Collins Radio Company but produced under contract by several manufacturers. Some maintain it's the greatest receiver ever made.

Only in recent years have better receivers overtaken the R-390's primacy. It can be like a search for the Holy Grail to find one of these old beasts, since those who have them seldom part with them, and no more are being made. These receivers require care and feeding, including scrounging for replacement vacuum tubes in a solid-state world. The R-390 is not a good choice for beginners. However, you might eventually find yourself under one's spell.

The Collins R-390A is a highly regarded tube receiver

Keep a sense of perspective

In your reading, you will discover there is little your fellow hobbyists enjoy more than swapping tales about their favorite equipment. You will get a notion of "state of the art", but please keep some perspective. As a beginner you can relax and enjoy listening on your present receiver for a long time before you "need" to move on to more advanced equipment. Don't be overly impressed with the loggings by someone using a megabuck receiver with half a dozen advanced accessories.

Remember, my friend, even this guy was a beginner at one time. All that expensive equipment might bring in a few more signals but I'll bet this person would trade it all in if the thrill of the first 100 loggings could be experienced all over again. And, as you will discover as you move along in the hobby, skill can take the place of money any day of the week. You doubt me? I could point you in the direction of a guy who does all of his AM monitoring using a simple home-built crystal radio. Probably the most basic radio receiver you can construct for literally pocket change. He has over 30 states and a few foreign contacts in his log book.

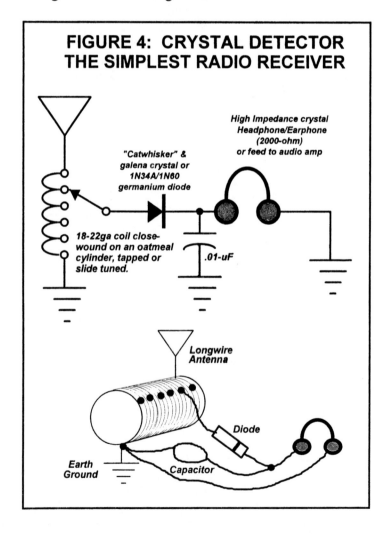

FIGURE 4: CRYSTAL DETECTOR THE SIMPLEST RADIO RECEIVER

9

Mediumwave antennas

Since you have come this far, I can begin to make a few assumptions about you. It would be easy to say that you are probably a very curious individual. Since you are so curious it might also be reasonable to assume that sometime in your life you may have taken the back off of a radio receiver and looked inside. If you never have done this I am also obligated to remind you (before your curiosity gets the better of you in the same way it did the cat) that the voltages present within many electrical devices are sufficiently dangerous that you should avoid contact with things you do not understand. The warning found on the back of most electrical gear that reads NO USER SERVICEABLE PARTS INSIDE really means business unless you have taken the time to learn and understand how dangerous electricity can be and you have some background in basic electronics. Having said this, if down the road you want to learn how to do these things, you may want to invest some of your hobby time in developing these technical abilities. Once you have acquired the requisite skill, experimenting with electronics is gobs of fun. But please, find yourself competent, certified instruction. I want you around to enjoy this hobby for many years to come.

Ferrite antennas

Anyway, inside most AM receivers you will find an antenna. You may not recognize it as such, but it is there. The antenna will usually appear as a short black rod with some wire coiled around it. This is commonly referred to as a **ferrite rod antenna** although some old-timers might call it a "loopstick." If you are looking at the insides of a common AM/FM portable, the **whip antenna** that sticks outside of the case (and always breaks off when you least expect it) has nothing to do with AM signal reception. The whip is usually just for FM.

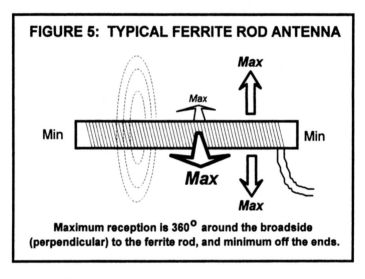

FIGURE 5: TYPICAL FERRITE ROD ANTENNA

Min   Min

Max

Max

Max

Max

Maximum reception is 360° around the broadside (perpendicular) to the ferrite rod, and minimum off the ends.

As you have already discovered from some of the activities and experiments we have discussed a while back, this ferrite rod antenna exhibits a *directional* reception pattern. This pattern is broadside to the length of the rod. Ferrite antennas are usually oriented along the length of the receiver's case. In effect, you should be able to orient the back or front of your receiver in the direction of the station you are trying to hear. Pointing the ends of the rod antenna toward an incoming signal will result in a "null," making that signal more difficult to hear than any other signal coming in along the broadside of the rod. You probably have already taken advantage of this fact to null and peak signals to allow you to discern more than one station on a given frequency. Just this amount of knowledge gives you the potential ability to double the number of stations you can hear with no further action taken on your part other than listening and logging.

Loop antennas

But at some point, you will probably want to step up in antenna performance. Take another look at the picture of the ferrite rod antenna in *Figure 5* (or at an actual one if you have progressed to that stage of curiosity). Notice that what you have is a tiny coil of wire wound around the rod making a **loop antenna**. This is a perfectly efficient antenna for common listening practices. Ah, but radio monitoring is an uncommon practice. We want more performance to hear those uncommon and even rare signals. Could bigger be better? You bet! How about if we construct a much larger loop? This practice of loop antenna construction is one of the most fun aspects of the AM monitoring art. As you read through AM monitoring club publications you will discover that hobbyists have developed dozens of loop designs. And why not? Loops are inexpensive to build and fun to experiment with. As you progress in the hobby, you will no doubt throw your own design ideas into the discussion. But for now, consider one of the more basic designs.

The basic "2 foot" box loop antenna is very simple to construct out of mostly found materials, some wire, and one store-bought electrical component. Let's take a look at this particular loop design as a point of departure for future experimenting. This antenna design is as old as radio itself. If you have ever seen an exhibit of antique radio equipment, you have probably run across some variation of this design. Even in the high-technology world we live in today, this most basic antenna still holds its own.

What you see in *Figure 6* is a very basic variation on the loop design. Essentially, you create a frame out of wood, plywood, plastic, PVC pipe, or just about any nonmetallic material you can form into a cross pattern that is 24 inches across. The width or depth of these "spreaders" needs to be at least 2 inches wide, but give yourself some room to work with. The material needs to be strong enough to stand up to being wrapped with wire, but plan to keep things light enough to allow the loop to be handled easily. You will find, as you encounter other loop designs, that loop frame construction is an area of high creativity. Try whatever you have around the house before you go out to the building supplies store. Keeping costs down can be part of the fun of radio monitoring. Think of the sense of accomplishment when you succeed in logging a station with a freebie antenna, when you know folks who struggled to hear the same signal with a megabuck commercially produced loop. Even with all store-bought materials, a fully functional loop can be built for well under 20 dollars.

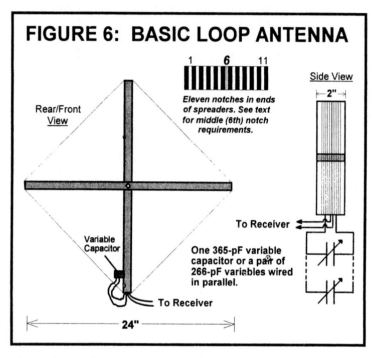

FIGURE 6: BASIC LOOP ANTENNA

1   6   11

Eleven notches in ends of spreaders. See text for middle (6th) notch requirements.

Side View

Rear/Front View

Variable Capacitor

One 365-pF variable capacitor or a pair of 266-pF variables wired in parallel.

To Receiver

To Receiver

24"

Along the width of the four edges of the spreaders make 11 notches spaced about 1/8" apart. These grooves will allow the wire to sit uniformly when you form the loop. On the bottom spreader, nail in one small nail each at the first and last (eleventh) notch. Also at the middle (sixth) notch, nail in two small tacks. These tacks will serve as fastening points for the wire you are about to wind.

Using a relatively light-gauge insulated wire (something around #24–#26 should do nicely), wind a loop of wire around the spreader structure. You will begin at the nail at the bottom first notch by fastening the wire around the nail so as to leave about a foot of wire hanging loose, while the rest of the wire goes around and around the loop structure. Continue winding the loop uniformly around the structure until you come to the sixth course of notches. Skip over the sixth notches all around the form, continue to wind the loop from the seventh notch, and carry on until you finish off at the nail at the eleventh bottom notch. Again, be sure to leave about a foot of wire to work with at the end. The two ends you have hanging down at the bottom are then attached to the two sides of a 365 pf (picofarad) variable capacitor. This is essentially the same part that is used in the tuning mechanism of most AM/FM receivers. It can be found at any electronics store that sells individual components. If you want to go really low budget you can salvage this part out of a broken inexpensive AM pocket radio or portable. At the most this part should cost between $1 and $3 new and will probably be the biggest expense of the entire loop project. If you cannot find a single capacitor that has the 365 pf value, you can gang two 266 pf variables together in parallel. They are a more common value and easier to find. This part serves to "tune" the ten-turn loop of wire, allowing you to peak signals coming in on a specific frequency. Mount the capacitor somewhere convenient on the bottom spreader arm. Try to keep the wires coming away from the loop as short as possible. I just recommended a foot of wire to give you enough to work with and even make a mistake or two.

Before we go on to make use of those sixth notches, I want you to try something with the loop as far as it is already built. Take your AM receiver or any small portable and tune it to a weaker station. Now, with the loop oriented so that its windings are pointed in the same direction as the internal ferrite rod antenna of the receiver, bring the loop as close to the receiver as you can. If your spreader structure allows for it, insert the receiver within the loop structure itself. You should hear at least a slight increase in the incoming signal. Now fiddle with that variable capacitor on the loop. You should be able to peak the signal by adjusting the device. Pretty neat, huh? What you have just experienced is an electrical principle called "inductive coupling." As you can see, you can already make use of the loop without even physically connecting it to the receiver. So in this way it is possible to use a loop antenna to improve performance of even inexpensive receivers that have no provision for external antenna connections. This is another way to keep the costs of radio monitoring down by using what you already have instead of running out to buy a new receiver.

But let's say you have, or have decided to purchase, a receiver with an external antenna connection. This is where those sixth notches come into play. Using the same process that you used to construct the ten-turn loop, make a one-turn loop using the two nails and the notches at the sixth notch position, essentially in the middle of the ten-turn loop. This one-turn loop also serves to inductively couple itself to the ten-turn loop; however, in this application, it allows you to bring wire or cable away from the loop to the antenna connections on the receiver. Your receiver will have either two screw posts or a jack for its external antenna. You will connect the one-turn loop so that it goes to these posts or to either side of the appropriate-sized plug for the jack. You can use two lengths of wire for this connection but a better application can be made using any low-cost "coaxial cable." This is the common form of cable that is used in most radio-frequency applications. You will recognize it by a single conductor running through the center of a tube of braided wire, separated by a plastic or foam dielectric material.

This type of cable is made in different sizes and ratings for transmitting applications. These ratings are not critical to this application so just pick out the cable that is relatively light and inexpensive. A common type you can use is RG-58 or its equivalent. Using this cable you must take care that you strip and connect the center conductor and braid to the two sides of the one-turn loop in such a way as to be sure that they do not come in contact with one another. One of the two antenna posts on your receiver will be marked *ground*. Connect the braid side of the cable coming away from the loop to this post and connect the center conductor to the other. If your application involves a jack and plug, usually the center conductor is wired to the tip of the plug and the braid would be connected to the outer or shield side of the plug. When wiring into a plug and jack situation you should always consult the manual that came with the receiver to assure proper connection.

Loops come in many variations. The most common designs are 2-foot and 4-foot sizes. Don't be afraid to try out new ideas. You can't really hurt anything and you can have a lot of fun.

10

Mediumwave accessories

The term *accessory* is kind of misleading. Usually it refers to anything beyond the receiver and antenna in a monitoring application. But the tendency is to think that it refers to things you don't really need. True, the folks in the business end of the radio hobby are ready to sell you all the bells and whistles that your wallet can tolerate. Many of these items are in fact things you can do without and still enjoy the hobby. Since this book is written for beginners, I plan to inform you in each section of those few "essential" accessories that are well worth the cost in dollars to improve your overall monitoring abilities.

Headphones

No matter what kind of radio monitoring you do, when it comes time to dig a weak signal out of the slop and static, nothing beats a good set of headphones. Headphones serve to cut down background noise and focus all of your listening abilities on the task at hand. First, you need to know a bit about your receiver. This information may appear on the receiver's case but you can always check the manual. If your receiver has a headphone jack, you need to know how it is wired so you can pick the right headphones for the job. Many AM/FM portables come wired with a headphone jack designed for use with common stereo headphones like those that folks wear with their radios when they are jogging. As a matter of fact, if your rig is so wired and you have a set of these lightweight headphones around, give them a try. They won't be quite as good as a set of communications headphones but, if you already own them, the price is right. If you plug in these headphones and you only hear sound out of one side or the other you have just discovered that your receiver's headphone jack is wired for monaural headphones. You can either rewire your existing headphones or buy a stereo-to-monaural adapter at any nearby audio/electronics store. If you find that the jack is in fact wired for the stereo headphones, you can start shopping around for a reasonably priced set of "full coverage" headphones. These are the more traditional design with the earcups that fully block out outside sounds. You don't need to go after the high dollar fancy audio headsets. You just don't need the expensive fidelity that you buy when you get these kind. You should be able to locate stereo headphones for our application for well under $15.

Back to those mono wired jacks. Equipment designed for communications monitoring is usually wired monaurally. With equipment so wired, you will want to shop around for headphones designed for the monitoring hobby. Headphones in this category will run from $35 through $80. Communications headphones have a "shaped" frequency response that works best with the human voice, and not the entire high-fidelity range that music oriented headphones have. As you use them, you will find these specialized headphones are worth the price. Also, a good set of headphones will stay with you through many different receivers. These monaurally wired headphones can be used with stereo jacked receivers, this time using a mono-to-stereo adapter.

Tape recorders

Tape recorders have countless uses in all types of radio monitoring. But more than any other practice, mediumwave monitoring benefits the most. Remember those local "graveyard" frequencies we talked about earlier? Tape recording and reviewing the multiple signals from such frequencies often makes the difference between a successful logging and a missed opportunity. Also, tape recording is a great way to provide confirmation of stations that you have heard when a station, for whatever reason, does not respond to your request for written verification.

Most modern receivers have at least a headphone jack. But be on the lookout for equipment that also has a separate tape recorder jack. This separate output provides the tape recorder with a constant output level so you don't have to worry about where your receiver audio gain (volume) is set during your listening session. If you have just a headphone jack, no problem. Just go to your local audio/electronics store and buy an appropriate sized "Y" adapter to turn your single jack into two jacks. To provide a signal that won't overload the tape recorder when using the output of the headphone jack, you can purchase an Attenuating Dubbing Cord at any audio/electronics store. Or, if you know your way around a soldering iron, you can roll your own out of a few simple parts and a bit of shielded audio cable.

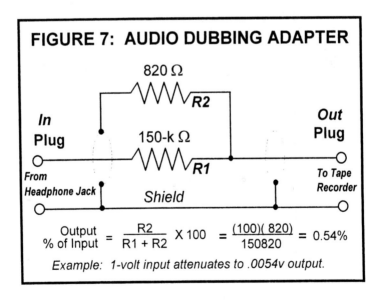

FIGURE 7: AUDIO DUBBING ADAPTER

$$\frac{\text{Output}}{\text{\% of Input}} = \frac{R2}{R1 + R2} \times 100 = \frac{(100)(820)}{150820} = 0.54\%$$

Example: 1-volt input attenuates to .0054v output.

All you need to do to make your own cable is to wire a 150 kΩ resistor into the center conductor of the shielded cable and a 820 ohm resistor from the shield to the center conductor. You may have to play with the resistor values to achieve optimum performance, but that is all part of the fun.

As tape recorders go, any low-cost standard cassette recorder with a jack for an external microphone will do nicely. Once again, you may already have such a device in your home. Remember, we are not looking for audiophile quality signals here. Actually low-cost recording equipment tends to emphasize the lower audio frequencies, which is just the ticket for radio monitoring. The only extra feature you may find helpful on any tape recorder you choose would be a tape counter to help you know where to look on particular tapes for certain information such as station identifications.

Maps

One other item should be considered as you start out in mediumwave monitoring. A detailed map of the United States or a good road atlas will go a long way in helping you figure out where stations are located, the time zones they are in, and the geographic conditions between you and the stations you are hearing. After a few months in this hobby, you will know the United States better than most graduating high school seniors.

11

Mediumwave propagation

Propagation ... **Now isn't that a strange word** to associate with electronics? We think of propagation in terms of breeding and spreading the word. In fact, you have to go down to about the fifth definition in most dictionaries to find that propagation is a physics term to describe "motion through a medium." Now we're on to something. The electromagnetic radio waves that carry information from transmitter to receiver move through the medium of the atmosphere. This medium has an affect on how far and well these signals travel. Further, the conditions in our atmosphere change often enough to make predictions concerning how a signal might behave a cross between science, luck, and some would even say magic. It's not that the affects of the atmosphere aren't understood. As a matter of fact in a few minutes I'm going to describe to you what goes on. The problem is these affects are constantly changing due to the impact of such phenomena as temperature, the location of the sun, the conditions on the surface of the sun, weather, and other factors. To further confuse the issue, poor conditions for the transmission of radio signals at one group of frequencies may in fact enhance the propagation at another frequency segment.

As a beginner, you will have to accept that propagation is a science that has aspects that elude even the most experienced radio monitors. The best we can hope for is to understand a bit about how propagation works so that we can make good decisions about when and where to listen. You must also realize that, at times, the propagation conditions will conspire against your listening habits and opportunities. That is just one of the challenges of this kind of listening that makes the rare catch that much more rewarding when it occurs. The most important fact you can come away with is that propagation is like the weather in one important way. If you don't like the way things are now, stick around, they're going to change.

As we look at the different types of monitoring; mediumwave, shortwave, and VHF/UHF, you will see that propagation differs from one to the next. For now we will concentrate on helping the novice AM broadcast band monitor to listen smarter and make more loggings. You will find, as you spend more time at the receiver, that you will develop a good practical understanding of how propagation works both for and against your particular listening goals.

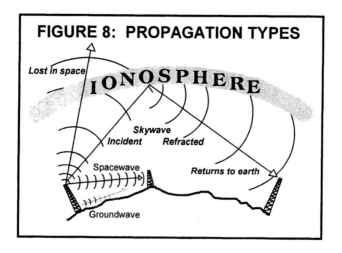

FIGURE 8: PROPAGATION TYPES

There are about half a dozen major propagation types that radio monitors of various stripes worry about. However, in the world of mediumwave listening there are really only two that normally come into play. This is yet another reason why I maintain that AM Broadcast Band monitoring is the ideal place for a beginner. These two types are known as **groundwave** and **skywave** propagation.

Groundwave propagation

You are probably familiar with the location of some AM broadcast stations in your area. When you are in a position to see the radio station's tower from your location and receive its signal on your home or car receiver, that is referred to as **line of sight** propagation. But how do you receive a station that's beyond line of sight to your location? Most stations within a few hundred miles of your location (but not in Line of Sight) are received thanks to the affects of groundwave propagation. The station's signal is literally conducted along the surface of the earth.

There are a number of issues involving Mother Nature that have a direct effect on how well a groundwave signal travels. One issue is the mineral content of the ground over which the signal travels. Ground conductivity varies greatly from place to place around the country. But in general, ground conductivity is poorer toward the coasts of the United States and better toward its center. Does this mean that folks on the East and West coasts should not take up mediumwave monitoring? No, because we are only talking about one general factor. Ground conductivity is also greatly affected by seasonal and immediate changes in the moisture content of the soil over which the signals travel. Keep an eye on the regional weather forecasts. If, for example a large rain storm sweeps across your area and several nearby states, don't be too surprised to find improved groundwave propagation when you settle in to listen after the storms pass. Also groundwave propagation over large lakes and bays will prove excellent. Actually, groundwave reception over sea water is many times better than it is for the same distance over any land, regardless of how good its ground conductivity may be. Maybe we should refer to this as "sea wave" propagation. It might be worth taking a trip to such locations to pick up a few new stations. Nobody said you have to do your listening at home!

Alas, Mother Nature gives but she also takes away. The very ground that those radio signals travel over also gradually absorbs the signal along the way, causing it to weaken until at some point it is no longer discernible. Also, even though signals travel along the earth, there is a noticeable blocking effect caused by hills and mountains.

Groundwave effects are more prevalent at the lower end of the band (530 kHz), and they decrease as you move up to the top of the band (1610 kHz). Groundwave is the most dependable form of propagation during daylight hours but its affects can also be noticed at night. You will discover that groundwave propagation is relatively dependable up through about 150–200 miles, although it can be noted to significantly longer distances due to seasonal and regional considerations. In general, you will find groundwave propagation to be better in winter than in summer. Groundwave propagation will easily lead you to your first long-distance loggings. But it's only the beginning: even longer distances are possible.

Skywave propagation

Skywave propagation is the primary means by which distant stations can be heard at your receiver. It is the key to filling a log book with signals that non-monitors would think impossible.

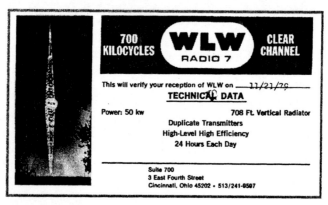

Skywave propagation lets WLW be heard on both coasts of North America

The earth is surrounded by a ball of gases that extends up more than 1200 kilometers (although I would not want to try breathing up that high). The area between 60 and 1200 kilometers (km) is generally considered the **ionosphere**. You probably learned about it in some science class, but had no notion that it would be important someday. The ionosphere is very important if you plan to fully enjoy the radio monitoring hobby. The gases between these altitudes become "ionized" by the **ultraviolet radiation** from the sun. The more radiation, the more ionization. So it's easy to see that the ionosphere will be more densely ionized over the part of the world that is exposed to sunlight at a given time. Now think for a moment, you have some very important information. If the sun affects the ionosphere, then the sun affects what you can hear on your receiver.

Skywave propagation is apparent when signals literally bounce off of the ionosphere and carom back to earth. Technically this phenomenon is known as **refraction** but it is referred to as **skip** because the RF signals skip along the ionosphere similarly to the way a stone skips along a pond's surface. Sounds simple enough, right? But there's more to the story. The tricky part is that the ionosphere goes through a series of changes that constantly affect how this refraction occurs. These changes determine how well a signal is refracted, the angle at which the refraction occurs (think of a "bank shot" in the game of pool), and even the time of day that the refraction will take place successfully. These factors are tied directly to the earth's relationship with the sun.

Ionization density

Now comes the interesting part. If the ionosphere is very dense, too much of the transmitted signal will be absorbed and not enough signal will get back to earth to be heard acceptably. On the other hand, if the ionization is not dense enough, the transmitted signal will not refract as desired, instead it heads off into space. Toss in the fact that the altitude at which refraction occurs also varies with ionization density and you can see why the science of propagation may seem about as accurate as tabloid newspaper astrology.

With almost a century of professional and amateur propagation prediction under our belts, certain discoveries have allowed radio people to make good predictions about when and where to listen for just about anything. Over the years, students of the radio art have come to know patterns of ionospheric density that give us a clearer understanding of propagation. You can imagine these patterns as layers of atmosphere stacked one on top of the other. The layer closest to earth is designated by convention as the "A" layer and as we move out from the earth we would encounter layers "B," "C," "D," "E," and "F." For the purposes of mediumwave monitoring, we only need to worry about the "D" and "F" layers. Other layers will have meaning as we look later at shortwave and VHF/UHF monitoring.

If you wonder why you can only hear some of your favorite distant monitoring targets in the evening and not at noon you can blame the ''D'' layer. This region is usually between 50 and 100 km above the earth. Its ionization is very low at night but it becomes very densely ionized during daylight. So dense in fact that it absorbs any signals below 7000 kHz, effectively blocking all long-distance mediumwave communication. Remember how I said this ionization process was related to the sun? Intense solar flares highly ionize the ''D'' layer so that it can blank out all radio frequencies, leaving only local groundwave communication possible. So the best time to listen for far-away broadcasters is during the hours when both your location and that station's location are in relative darkness, thereby minimizing the negative effects of "D" layer absorption.

Between 160 and 320 km, the ''F'' layer is the last layer of the ionosphere that affects radio waves. Actually I should say layers, because during daylight, "F" splits into two distinct layers at about 200 km. Propagation pundits refer to these layers as F1 and F2. But for our purposes in the mediumwave monitoring realm, you can just consider the "F" layer, because the time of evening that this layer becomes important to mediumwave monitors occurs after the F1 and F2 layers have combined into a single layer at around 250 km above the earth. The ability of the ''F'' layer to refract signals depends directly upon the density of the layer's ionization. If the density is too high, a greater degree of absorption occurs, weakening the signals along their path, not unlike what happens in the ''D'' layer during the day. The ''F'' layer is denser in summer than in winter.

After the sun sets, the "D" layer no longer blocks long range communication. The ''F'' layer then allows mediumwave signals to refract over great distances. This process is predictable enough to allow the radio monitor to take advantage of these changes. The plus for the radio monitor is that variations in refraction angles and even multiple hop signals can allow hearing stations at distances vastly greater than their intended target audiences. By monitoring regular seasonal variables and patterns of solar activity a dedicated monitor can catch signals not hearable under normal conditions.

This primer represents a brief insight to how propagation relates to your listening practices. Entire books have been written on the subject. However, this little bit of knowledge will allow you to begin to have great accomplishments as a mediumwave radio monitor.

12

Mediumwave monitoring techniques

Anybody can sit and listen to a broadcast band receiver. People do it every day without giving it much thought. What sets the mediumwave radio monitor apart (and makes that individual successful at the craft) is the ability to listen *systematically*. On the AM broadcast band, there are literally thousands of potential signals knocking at your door. The key to adding as many of these signals as possible to your log is a combination of organized listening and a good knowledge of what there is to hear.

Three-pass band study

A good start out is a "three pass" study of the band as it appears to you at this point in your monitoring career. This will serve to make you sufficiently familiar with the listening environment present at your location. Remember that in the AM band, you must contend with both changes in propagation and changes in station operating practices throughout each 24-hour period. A handle on the regular changes in the listening environment will alert you to opportunity when the more unusual events occur. With this knowledge under your belt and in your log book, you'll be better able to spot those rare catches among the more familiar signals.

Your first pass at listening through the band should be in daylight. This will give a good idea of all the stations that can be heard from your location that are operating at their usual full power and coming in by way of groundwave propagation. Most of these stations will be familiar, since their programming has probably been known to you long before you caught the radio monitoring bug. In theory this could be as many as 100 stations, but more likely you will note between 25 and 50. Make a list including each station's frequency, call letters (KABC, WIBG, etc.), and make a note as to the station's format (e.g., Rock Music, Country Music, News, Religious, etc.). Do this from one end of the band to the other. Don't dig too deep in the static for the tough ones right now. You're just establishing a baseline by which you can grab those hard ones later.

The second pass should be done after full darkness but before local midnight. Many of the signals that you discovered on the first pass will still show up, but a lot of new and different signals will share space with them. This is the result of both the stations' regular sign-offs and low power periods. You will also experience to some degree the normal effects of skywave propagation. One trip through the band after dark and you'll have an idea about why this form of radio monitoring is so interesting. Again, make a second list with the same information as you did on the first with one additional bit of data. If a station's power level has changed noticeably from your daylight examination, make note of this fact on the list as well. Here you will begin to see some of the operating patterns of the stations that are more common to your nighttime band scanning.

Pass number three requires a bit more dedication to the task than the first two. Pass three should occur after midnight, preferably on a Sunday night/Monday morning. This is a time when stations often sign off for routine maintenance. Also, some stations close out their operating schedule at midnight. So what you hear "normally" will change once again in this time period. More so than any other group of radio hobbyists, mediumwave listeners appreciate the recent practice of juggling holidays to Mondays. It makes it easier to make such late night excursions into the radio monitoring world without drastically modifying their lifestyle. Once again, this third list you will make as a result of this bandscan should give you even more information that will help you in your listening down the line.

You now have three working documents to keep handy at your listening post to help jog your memory about what is where as you go digging through the bands in search of new and different signals. But perhaps without noticing it, you have probably logged between 75 and 100 stations over a span of between at least 10 and 15 states. You have already become an accomplished monitor well on the way to greater listening success. And I'll even be willing to guess that you had quite a bit of fun along the way.

Armed with the above information, it's time to go hunting. At this point you already have enough monitoring savvy to develop your own listening style. For instance, the times of the day or night you can devote to monitoring might differ from the times I set aside to listen. Your available listening time will probably dictate more about how to pursue new loggings than any other factor. Just remember that the best time to listen for most folks is to listen whenever you can. Your rewards and successes might be different from someone who monitors at different hours, but they will be just as sweet.

Targets of opportunity

Now that you are monitoring in earnest, a few tricks may be in order. One strategy is to just tune up and down the dial for targets of opportunity: i.e., those stations that pop up that didn't come across when you made your original passes to generate your information lists. This is probably the most common practice early in a person's monitoring career, and will yield good results for a time.

Specific frequency monitoring

Others like to camp on a particular frequency for a few hours, sometimes for a few days, and just try to wring as many signals out of it as they can. This practice is very popular when you are dealing with a regional frequency with quite a few signals coming through or on the graveyard frequencies: 1230, 1240, 1340, 1400, 1450, and 1490 kHz. This style requires a lot of patience but it will also serve to increase your loggings if you are willing to be persistent.

Map listening

Maps are great planning tools for listening. Get a detailed map of your county or region. Your site should be close to the center. Draw concentric circles around your site at radii of 150, 250, 750, and 1000 miles. Using the resources we will talk about in Chapters 14 and 15 of this section, you will make notes of stations in the areas between these mile rings. Then, based on propagation conditions and station operating practices, try to catch all stations within each ring.

Clear channel, regional, and graveyard frequencies

Another tactic is to concentrate on Clear Channel stations until these possibilities are exhausted. Then move on to the regional stations. After the regionals run out of steam, focus on the local "graveyards." By the time you cycle through this process, there will be enough changes in operations, call signs, and seasonal propagation to warrant repeating the process again. One of the few constants in the AM broadcast band monitoring world is change.

Club loggings

As you become aware of and familiar with the club publications that are available to assist the AM monitor, you will discover yet another way to skin the cat. Club bulletins often list the loggings of members. You may note that members who are in your region are having particular successes at going after certain stations at certain times. This might serve as a guideline for your listening pattern to grab those signals. Also club bulletins publish notices of station's "off air" maintenance, testing, and construction periods. All of these tidbits of information can lead you in the direction of new signals.

Hit lists

Some people develop "hit lists." These are lists of stations that they have particularly targeted as goals. This could be a list of the two or three most likely stations in each of the 50 states. This is a way of charting your ability to "hear" your way across the country.

Antenna skills

Don't forget to take advantage of your receiver's antenna, be it internal or external. No frequency is exhausted until you have made an effort to null out those stations you can hear easily, in an attempt to catch something coming in from another direction. Making full use of these directional receiving patterns will become one of the most important monitoring skills you can develop.

Experiment!

No one method has ever demonstrated any clear advantage over any other. Any technique you choose will only be as effective as you make it for yourself by way of your own patience and tenacity. My suggestion is you give each of these methods a fair trial until one or two of the techniques bears fruit for you. Down the road, if your success rate slacks off, you can always try some of the other techniques that you had earlier set aside. You may even come up with a notion or two of your own that helps you get results. There's no law that says you can't experiment and even play. The only rules here are that you enjoy yourself and you enjoy what your doing. This is a hobby, it's okay for it to be fun!

Sunrise and sunset opportunities

Dedicated AM-broadcast monitors have a few more tricks up their sleeves to wring even more stations out of those standard 10 kHz channels. Remember that the Federal Communications Commission establishes operating rules for each station that it licenses. These rules become interesting and useful to the radio monitor at sunrise and sunset. Many stations are expected to make significant adjustments in their sign-on, sign-off times as well as in their power level in relation to their local sunrise and sunset times. These adjustments are designed to avoid interference between stations as groundwave propagation gives way to skywave propagation. Many "daytime only" stations also apply to the FCC for permission to operate at a relatively low power from 6 AM until local sunrise and until 6 PM after local sunset to allow for program continuity at those times of the year when the sun and the clock are not cooperating with one another. The two twilight periods of each day offer the dedicated radio monitor the chance to catch stations that are signing on or signing off.

Sunset Monitoring

Let's look at sunset listening first. Generally the best time to go sunset station hunting is from around one hour before your own local sunset until about two hours after. As you may remember from one of your past science classes, the angle that the sun strikes the earth has an effect on the seasons. This phenomenon is why the length of the day and the times of sunset and sunrise vary throughout the year. It also affects the angle of the sunset terminator as it travels across the earth. Therefore, from season to season, you will find different stations in the twilight "footprint." So sunset listening should fit into your regular monitoring plans throughout the year.

Of course you will be keen to hear new and different signals that come your way, but you will want to watch for daytime stations giving their "sign-off" announcements. You will take advantage of the FCC's policy of assigning "official" sunset times based upon the time of a station's local sunset on the 15th of each month. Stations must obey the FCC, but good old Mother Earth moves to her own rhythms, and creates a situation in which stations will actually broadcast *after* their local sunset during the first half of each month in the spring and during the last half of each month during the fall. This little bureaucratic decree allows you to catch stations that would not otherwise be possible. During these windows of opportunity, you can camp your receiver on any frequency that has multiple daytime stations and just wait. You will hear stations signing off as often as every fifteen minutes. It is not unusual to log up to eight stations in a row on one frequency in this manner. It helps to have a tape recorder to catch the station sign-offs for later evaluation.

Stations signing off will not be the only signals heard when sunset monitoring. Throughout the year, different regions become open. Again, this is due to the changing angle of the sunset terminator brought about by changing angles of the sun's rays hitting the earth. Propagation is directly affected by this phenomenon. There will always be unique shifts in what is heard as atmospheric absorption of signals gives way to skywave propagation. Sunset listening favors eastern monitors because of the direction of sun travel across the land. Western listeners can still take advantage of this practice but will often find better luck at sunrise.

Sunrise Monitoring

Sunrise listening rewards the dedicated monitor. Some stations use the FCC's permission for Pre-Sunrise operation at low power until the appointed time for full power. Local sunrise is determined for bureaucratic purposes on the 15th of each month. Watch Regional and Clear Channel frequencies for station sign-on signals.

You can take advantage of a basic human quirk when monitoring sign-on signals. See, everyone is rusty early in the morning, including station operators. WZZZ may be scheduled to fire up with a pre-sunrise power authorization of 500 watts at 5:45 AM, but the person on the switch takes longer to get through his check list that morning so he doesn't actually get anything going out to the antenna until 5:47 AM. Meanwhile this break allows weak station WAAA's sign-on identification at 5:45 to be heard before it's wiped out by WZZZ. The same thing can happen again at around 6 AM if the station operator is a few minutes slow at kicking the station up to its full authorized power. Another station that is on time with raising its power level might get through to you and your log book. You have to be quick and pay attention, no small trick if you are not a morning person, but sunrise monitors get opportunities like this every day.

There are a few practical tips to early morning monitoring. Have your listening post all set up and ready the night before so you don't have to waste time pulling yourself together like the station operator of whom you hope to take advantage. Get up early enough to "wake up." Take a shower or get a couple of cups of coffee into your system so that you are alert to what you might hear. If you have a tape recorder have it running throughout the session because things will happen quickly and you won't want to miss anything.

Emergency monitoring

There is another time that you can have unique opportunities to hear stations that might normally not show up to you. In times of emergency or disaster, stations can be authorized by the FCC to operate at full power beyond their normal sign-off or power-reduction schedule. Floods, tornadoes, hurricanes, or other unusual occurrences can result in catching stations from the area of the fiasco by way of skywave propagation when normally they would be lost in the static. Keep an eye on the news and weather around the country and you might just turn up a few extra special loggings.

Foreign stations

Foreign radio stations will become prominent as you advance in monitoring. First you will hear stations from Canada, Cuba, Mexico, the Caribbean, and Latin America. Most of these will be on the same frequencies that you hear domestic stations. Most stations in the Western Hemisphere operate with the same 10 kHz spacing as in the United States. In Europe and Asia, you will encounter "split" frequency signals. These are stations that broadcast in countries that set their stations only 9 kHz apart. Skywave propagation allows the monitor to hear quite a few of these signals in between the normal domestic signals. East coast monitors have grand opportunities for transatlantic signals, while folks in the west get the transpacifics.

Summary

Keep in mind we have just scratched the surface of the skills used by mediumwave monitors. As you grow into the hobby you will learn about other strategies and you may even come up with a trick or two of your own. The diversity of signals that can be found on the AM broadcast band will keep you challenged for a long time to come. You will know the thrill of hearing things on "Mom's kitchen table radio" that most folks never even dreamed were possible. The important thing to remember at this stage of the game is that it is okay to be a beginner. Use this time to experience new things and develop skills and rhythms that suit you and your personal goals and desires. The only person you have to impress is yourself.

13

Time

The subject of time constantly comes up in the radio hobby. Most folks get along in the world by looking at their watch or the clock on the wall and that's it. Radio monitors find that time is a very flexible concept. When you deal with a signal that comes from some distance, you have to face the fact that you and the transmitting station do not exist in the same time zone. Keeping track of these differences, coupled with the interjection of Daylight Savings Time, can be confusing. Basically, in the mediumwave monitoring world, you can keep things logical by keeping "two sets of books."

As you document catches in a log, make two time entries. One entry will be for the time at your monitoring site and the other for the local time at the transmitter site. Clearly note Standard Time or Daylight Savings Time. Most folks record time in the 24 hour format to avoid confusion. This is where time is in the traditional format from midnight through noon, and then 12 is added to each of the second twelve hours, making 1:00 PM = 1300, 2:00 PM = 1400, 8:00 PM = 2000, etc. At this point you are simply applying the 24-hour format to traditional time. This is not to be confused with Greenwich Meridian Time (GMT) or Universal Coordinated Time (UTC). These methods of time-keeping are important when keeping track of overseas signals. We will talk more about these time-keeping methods when we get into working with shortwave.

The reason you need to keep track of time at both yours and the station's location is so you can give a logical and easily understood signal report to the station. This is often the key to successful confirmation of your catch. If you send a signal report to a station three time zones away, reporting only in your own local time, the person reading your report might not take the time or understand the need to make a time conversion. In such a case, your report might appear to be in error and the station may not send out a confirmation.

A 24-hour clock is a useful tool in the monitoring shack
(Photo courtesy of MFJ Enterprises, Inc.)

An atlas or map of the United States will indicate the various time zones that work their way across our country. Remember that 23 states fall under two time zones, so you will need to pay close attention to the transmitting station's location on your map. Also remember that Arizona, Hawaii, and part of Indiana have exempted themselves from the Daylight Savings Time policy by state laws.

> **"Time checks" can be useful in identifying stations because they are given in the distant station's "local" time.**

This may sound a bit confusing but you'll get the hang of it in no time. Actually you have help along the way if you keep your ears open. Most stations give "time checks" around the same time they give out their station identification report so you should often be able to get the transmitting station's local time in this way. As you grasp these concepts of time, you become more aware that the world around you is a much larger and more interesting place to live in.

14

Mediumwave clubs and organizations

One of the long-held myths of the radio monitoring hobby is that it is an isolated practice. True, most folks do a lot of their listening on their own. But you will soon discover that you can't move ahead in this hobby without some contact with the outside world. In every area of the radio-monitoring hobby, clubs and organizations have come into being to allow monitors to exchange information, learn new skills, receive information concerning the best equipment, and share what they are hearing. Mediumwave monitoring is no exception to this rule. AM broadcast band clubs have been around almost as long as the AM broadcast band itself. You will find that club membership can be one of the best resources for learning about the hobby and growing in your understanding and enjoyment of the radio-monitoring art. Two major clubs currently have a great deal to offer the beginning mediumwave monitor.

DX News

♦Serving DX'ers since 1933♦

Volume 63, No. 4 - October 23, 1995 (ISSN 0737-1659)

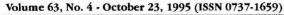

CPC Test Calendar

WDJL	1000	Oct. 23, 1995	0200-0230
WAMR	1320	Oct. 23, 1995	0300-0400
KREW	1210	Oct. 28, 1995	0330-0400
KFAY	1030	Oct. 30, 1995	0300-0400
KVON	1440	Nov. 5, 1995	0300-0330
WSEZ	1560	Nov. 6, 1995	0100-0130
WVAM	1430	Nov. 13, 1995	0130-0200
KNXN	1470	Nov. 13, 1995	0200-0230
KATD	990	Nov. 13, 1995	0300-0330
WKBO	1190	Nov. 20, 1995	0000-0030
WOON	1240	Nov. 20, 1995	0100-0200
KLCL	1470	Nov. 20, 1995	0330-0400
WJYM	730	Nov. 27, 1995	0000-0030
WPWA	1590	Dec. 11, 1995	0000-0030
WCGW	770	Dec. 18, 1995	0100-0130

From the publisher ... Jerry Starr is back from bringing rain to Florida, or so he says, and so is his "AM Switch". But Jim Renfrew is out with the boys, DX'ing exotic stations, and so he plans to return "IDXD" to these pages next week.

Why was #3 late? Well, when I take it to the printer at about 6:45 am Monday, the boss isn't in. So I put the pasteup where they tell me to. And that was a mistake, as that issue didn't get noticed nor printed until Friday that week. We think we've got our procedure straight, so look for it to be on time from now on.

Want to play in the FCC database, online? Rick Robinson notes that it's available at

http://radio.aiss.uiuc.edu/~rrb/fccdb.HTM

but he says that a graphics based Web progra, such as Netscape, Mosaid, or the AOL or Compuserve netware is needed.

Speaking of online communication, don't forget to send your E-mail address to Paul Mount at PMountNJ@AOL.com so that he can update the NRC E-mail list and publish it soon.

Keep an ear on 1660 from now on. Mike Hardester says that several people have reported hearing what they think is WJDM, Elizabeth, NJ, although one DX'er heard programming on 1640 from his location in Missouri. And NRC's Al Merriman heard rock music and an OC on 1660 on 10/10, and Dave Braun heard the same programming on 10/13 at 2035, but it was gone by 2100.

Ken Chatterton is compiling the verie signers list, and if you have information on verie signers that don't fit Ken MacHarg's requirements for his "Confirmed DX'er" column, send them directly to Chatterton. He'll pass along any pertinent to MacHarg, anyway, if you're not sure whether or not your list should go to Ken or Ken. (?!)

DXChange ... Denis Picard - 3830 St-Germain - Trois-Rivières-Ouest - PQ G8Y 6L8, 819-379-5647 is selling a Quantum loop for $135 CAN (approximately $100 U$).

Welcome to these new members ... Gary Durbin, Massillon, OH; Frank Latos, Farmington Hills, MI.

DXN Publishing Schedule, Volume 63

#	Deadline	/Pub. Date	#	Deadline	/Pub. Date
5.	Oct. 20	Oct. 30	18.	Jan. 26	Feb. 5
6.	Oct. 28	Nov. 6	19.	Feb. 2	Feb. 12
7.	Nov. 3	Nov. 13	20.	Feb. 9	Feb. 19
8.	Nov. 10	Nov. 20	21.	Feb. 16	Feb. 26
9.	Nov. 17	Nov. 27	22.	Feb. 23	Mar. 4
10.	Nov. 24	Dec. 4	23.	Mar. 8	Mar. 18
11.	Dec. 1	Dec. 11	24.	Mar. 29	Apr. 8
12.	Dec. 8	Dec. 18	25.	Apr. 12	Apr. 22
13.	Dec. 15	Dec. 25	26.	Mar. 10	Mar. 20
14.	Dec. 29	Jan. 8	27.	June 7	June 17
15.	Jan. 5	Jan 15	28.	July 5	July 17
16.	Jan. 12	Jan. 22	29.	Aug. 2	Aug. 12
17.	Jan. 19	Jan. 29	30.	Sept. 6	Sept. 16

DX Time Machine

From the pages of *DX News*:

50 years ago ... from the October 27, 1945 DXN: Phil Nichols, E. Hartford, CT, reported that he had reports out to 21 stations, including Toulouse, Limoges, 2PK, TI4NRH, TIEP, WVUR Canal Zone, Sottens, Hilversu, KIY, KALL, and KSRO.

25 years ago ... from the October 31, 1970 DXN: Dave Schmidt was to join the engineering staff of WHOT as an audio engineer at the end of November, making a record of three active NRC members at one station, and six NRC'ers within a five-mile circle ... Paul Hart's "Patterns" article explained the technicalities of directional patterns of AM stations.

10 years ago ... from the October 28, 1985 DXN: Doug Beard, Springville, IA was "just about as mad" as he could get because the FCC's new agreement with Canada, which replaced NARBA, allowed for no protection of that portion of a Canadian clear-channel station's secondary (sky-wave) coverage area in the U. S.

The National Radio Club has been in operation since 1933

The National Radio Club

The National Radio Club (NRC) was founded in 1933 and has served mediumwave monitors exclusively throughout its history. The club publishes a newsletter 30 times yearly, called *DX News*, that contains recent loggings from members, station information, features on equipment and monitoring techniques, and even construction projects for antennas and accessories. Of additional use to the beginner is an exhaustive reprint service of articles from past issues that will help a newcomer find out information about particular areas of interest. The club also produces several publications that serve as essential tools for AM monitors. The club holds an annual convention where members can get together and share radio experiences each Labor Day weekend. More information about this club can be had by writing: National Radio Club: Paul Swearingen, Publisher, P.O. Box 5711, Topeka, KS 66605-0711. A sample bulletin and information about current membership rates can be had by sending first class postage to this address.

The International Radio Club of America

The International Radio Club Of America was formed in 1964 and is almost identical to the NRC in its offerings to the beginning hobbyist. Their publication *DX Monitor* comes out 34 times per year. With its west coast headquarters, this club might prove particularly useful to people on that side of the United States. More information is available by sending first class postage to International Radio Club of America, c/o Ralph Sanserino, P.O. Box 1831, Perris, CA 92572-1831.

Both clubs are often responsible for establishing **DX Tests** at stations around the country. These are special test programs set up specifically to help AM monitors hear stations from locations that might otherwise be difficult to catch. The clubs work along with station engineers to set up times when the stations might be best heard, such as when another stronger station is off the air for maintenance or construction. Information about these tests and when to listen can be found in the pages of the club magazines.

Most people who get serious about mediumwave listening will find it almost essential to belong to one or both of these clubs as resources for up to date information about the band and what people are hearing on it. You will also get the cumulative experience of all those other monitors past and present to help you along the way.

By the way, being a member of a club should entail more than just sending in your dues and reading the newsletters. You will find that all radio clubs encourage member participation. As you become involved with a club you will become aware of the organization's procedures for submitting your loggings and other information to the club journal. This is all part of the fun of being a member. Even though you are a beginner, the information you have will be of use to your fellow monitors.

15

Mediumwave publications

In addition to club publications you will find two national publications that provide information pertinent to mediumwave monitors as well as to other radio monitoring enthusiasts. These magazines can become springboards of knowledge to further your understanding of the hobby and all that is has to offer.

Monitoring Times

Monitoring Times magazine covers all aspects of radio monitoring. This monthly magazine was founded by Bob Grove, WA4PYQ, one of the most respected people in all of radio monitoring. Of particular interest to the AM broadcast hobbyist is the "American Bandscan" column devoted to the AM monitoring hobby. There are receiver reviews, antenna, and construction columns that often cover subjects pertinent to AM listening along with regular feature articles on general radio subjects. I've had the privilege of writing the *Monitoring Times* "Beginner's Corner" column since 1988. This magazine publishes a regular list of DX tests similar to those found in *DX News* and *DX Monitor*. Contact *Monitoring Times* at (704) 837-9200 or write to P.O. Box 98, 300 South Highway 64 West, Brasstown, NC 28902-0098.

"MT" and "Pop'Comm" are the leading hobby periodicals

Popular Communications

Another major radio monitoring hobby magazine is *Popular Communications* (also known as *PopComm*). This magazine was founded by another prominent person in the monitoring hobby, Tom Kneitel, K2AES, and is currently managed by Harold Ort, N2RLL. *PopComm* has feature articles and columns in the same vein as *Monitoring Times*. These columns include several areas that would be of particular interest to the AM broadcast band monitor. You can find *Popular Communications* on many newsstands or you can get more information by writing *Popular Communications*, 76 North Broadway, Hicksville, NY 11801.

Both of these magazines' contributing editorial staffs are made up of people who are recognized by their fellow hobbyists as experts in their areas of interest. A subscription to one or both of these magazines will provide the reader with the learning experiences that build upon the foundations that this book represents.

16

Mediumwave frequency and station resources

In mediumwave monitoring, you pretty much know the nature of the playing field. You just tune up and down from 530 through 1610 kHz (or 1700 kHz if your receiver covers the new band segment) and you will hear signals worth listening to. But there are thousands of signals crammed into those 108-plus channels. You're going to need some support to make sense out of it all.

AM Radio Log

One long-standing resource that is well respected with the hobby is the National Radio Club's *AM Radio Log*. This book contains a comprehensive study of all the stations licensed in the United States and Canada. You get complete information concerning the station's power level, antenna pattern, location, format, and mailing address.

The book is cross referenced by city, state, and call sign. Also useful for confirmation letters is a list of verification signer's names and titles. The reference is available to both club members and non-members. For more information you can send first class postage to Ken Chatterton, NRC Publications Manager, P.O. Box 164, Mannsville, NY 13661-0164.

World Radio TV Handbook

Another important resource for folks who are beginning to hear signals beyond the United States and Canada is *The World Radio TV Handbook* (WRTVH). This book, published annually, is essentially a comprehensive guide to all forms of broadcasting in the world. In it you will find complete information on foreign stations similar to what you find in the NRC's domestic-oriented *AM Radio Log*. This book will also be an important resource if your monitoring practices branch out into shortwave listening in the future. The *WRTVH* is available in many bookstores and is sold in North America through BPI Communications, 1515 Broadway, New York, NY 10036.

17

Mediumwave & computers

Okay, you bought this book to learn about the radio-monitoring hobby. If you wanted to learn about computers you probably had to reach over a rack full of computer books to even find this book. Still, I would be remiss if I didn't tell you that the radio-monitoring hobby has changed dramatically (and for the better) over the last dozen years or so because of the personal computer. Further, the future of the radio hobby is most definitely intertwined with computers. All modern receivers contain some form of microprocessor technology within their circuitry. Some receivers even go so far as to have data ports that allow them to be directly connected to a personal computer for control purposes. True, you can have a ball with the radio hobby and never go near a computer. There is no reason for you to run out and buy a computer unless you have that particular urge. But it is important for you to at least be aware of the existence of this symbiosis between radio and computers because, as you develop in the hobby, you will probably want to make some decisions about how this relationship will affect you.

But what if you are already a computer user? How can this "other" technology enhance your enjoyment of monitoring? The answer is simply, a lot! If you look at the more basic uses of personal computers, you will find a number of logical ways the computer can serve your pursuit of AM broadcast signals.

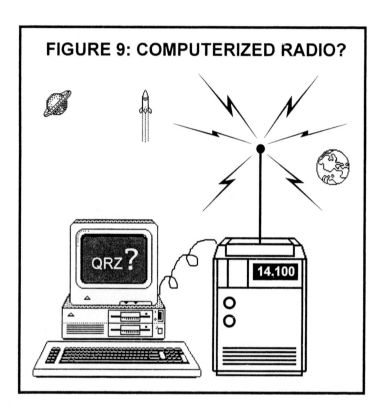

FIGURE 9: COMPUTERIZED RADIO?

Database management

We've been talking throughout this section about record keeping and maintaining an accurate log of what you have heard. Personal computer database management allows you to keep exhaustive records that can also be manipulated in various ways to provide you with additional perspectives. For instance, you may use a database program to create a log of all the stations you have been hearing. You would also include information in these records as to time, date, frequency, signal quality, program content, receiver and antenna information, and more. It is likely that you will be entering this data over time and in a consecutive manner as you hear each station. A few months go by and you want to check on your total station count for one of the "graveyard" frequencies. Instead of turning page after page in a handwritten log book to find all the appropriate entries (and possibly skipping some), just hit a few keys on the computer and the appropriate data is extracted for use. This information can even be written to an additional file. Speaking of files, a computer disk is small and easily copied. You could make a copy of your log onto a disk and store it in another location as a backup in case something happens to your original. For the typical radio monitor this process alone could protect years of painstaking documentation and research that would be all but impossible to recover if it were maintained as a written record. This applies to all your radio-monitoring documents that you put into a computer system.

> **Many of the pioneers of the "personal computer" movement were amateur radio operators. Wayne Green W2NSD1, publisher of *73 Amateur Radio Today*, was also the original publisher of *Byte*, one of the first computer magazines. Radio hobbyists remain at the forefront of this movement.**

Word processing

Word processing is the next place you will look to the personal computer for service. For many people, confirmation of the signals they have heard is an important part of the radio monitoring hobby. Many folks accomplish this by writing detailed letters to the various stations they have logged, outlining the programming they heard and the quality of the signal. In this letter they request that the station acknowledge the reception report with a card or a letter and they include return postage to accomplish this. The features common to word processing programs allow you to generate eye-catching and accurate letters. These letters can then be kept on file to allow you to send follow-up requests to stations that do not respond in a reasonable length of time. As you become more involved in the club aspects of the hobby, you will find many opportunities to correspond with fellow monitors. You will also find yourself submitting information to the club bulletins for publication. All these activities are made less of a chore through the use of the word-processing capabilities of any home computer.

The Internet

You probably have to be living in a cave to not know of the phenomenon of the "Internet." Computer users have this remarkable ability to communicate with one another over telephone lines via modem. Internet communication is world-wide in scope, instantaneous and inexpensive. It is no surprise that radio-monitoring hobbyists were some of the first to make use of this telecommunications ability to enhance their activities.

Many private and commercial computer bulletin board systems have areas dedicated to the various aspects of the radio hobby. The Internet provides many locations where monitors can gain access to computer programs and applications designed specifically for radio hobbyists. This ability to share information at the speed of light can greatly enhance the opportunity to catch signals that might otherwise be missed.

For instance, I recently checked the message section of a commercial on-line service that is set aside for radio monitoring enthusiasts. In this message area, someone posted information that a high-powered clear channel broadcaster near my home was going to be off the air late one evening for transmitter maintenance. That particular evening I tuned up on the proper frequency and logged three stations on that frequency and two more on an adjacent frequency. I never would have been able to hear those stations if I didn't have the knowledge supplied by way of the computer message base. Knowledge is always power in this hobby. The personal computer provides a sure pathway to more knowledge than you could develop by any number of other resources. All of the computer related activities discussed in this section can be performed with any common personal computer with readily available software. If you are currently a computer user, you may already have everything you need to get started using your system to enhance your radio monitoring activities.

18

Advancing your mediumwave skills

Some people use mediumwave as an inexpensive path of entry into the radio monitoring hobby. These folks often see AM broadcast listening as a springboard to other forms of monitoring. It can be that, but it can also be a wonderful lifelong monitoring pursuit in and of itself. Many people find the mediumwave band enjoyable enough that they make it the mainstay of their monitoring activity.

Like most activities, skill and expertise come from study and practice. Radio monitoring is no exception. Those you encounter at clubs or get-togethers who have huge signal totals and amazing rare catches in their collection are really no different from you, the beginner. The only thing the more advanced hobbyist has over the beginner is more time at the dials, more time listening, and more time learning. Successes as great or greater will also come to you over time.

Previous chapters discussed what's needed to start filling your mediumwave log book with those first AM-band catches that will be the foundation of your future "large signal totals and amazing rare catches." Now we turn to your future to discuss activities and ideas that will carry you into intermediate and advanced monitoring.

Before we get too deep, please *take time to be a beginner.* There's no clock to punch; no deadline to meet. Take time to savor your monitoring discoveries and accomplishments. The definition of a "rare catch" is any station you've never logged before. Later in your career, when you have to work a bit harder to snag the next new signal, you will fondly recall early experiences when everything was new. There is no shame as a beginner—every alleged expert began like you: plopped in front of a receiver wondering what's next.

Change is constant

The AM broadcast band is not a static environment populated with the same four or five signals day in and day out for years. Rather, the AM band is a dynamic environment. Stations routinely go on and off the air, or alter call signs; antenna patterns; transmitter power; operating procedures; and programming formats. Each change can be a new opportunity to log something new. Keeping current with change is one of the keys to success in mediumwave monitoring. Club publications are prime resources but there are others. A public library has great tools for mediumwave. One resource is the monthly *SRDS-Spot Radio Rates and Data Service,* from the Standard Rate and Data Service, with information on the operations of active AM stations. SDRS also has the names of people at the station. A reception report is much more likely to get noticed if addressed to an individual. The SRDS is updated monthly, so some of the information is more current than club publications.

> **"Library crawling" is a good habit for a radio monitor. I cannot stress the value of the public library enough. I've been in this hobby for more than half my life and I still find new resources every time I go to my local branch. Also, many libraries now provide public access to the Internet.**

Transmitter antenna patterns

Another area of advanced study is transmitter antenna patterns. After sunset, many stations switch to a directional pattern to serve specific areas and not interfere with other stations on the same frequency. A good working knowledge of night antenna patterns will give the dedicated monitor a better idea of what signals might have a chance of making a skywave skip or two in their direction. This information is available from various resources but the easiest for the beginning hobbyist to gain access to is the National Radio Club's *Night Pattern Book*. This publication gives you all the information and charts you will need to begin your own night pattern analysis for your particular listening location. More information on this publication and many others can be had by writing to the NRC Publications Center at P.O. Box 164, Mannsville, NY 13661-0164. Please include a first-class stamp with your request.

Monitoring conditions

Many hobbyists devote a similar study and experimentation to the **sunset and sunrise monitoring periods** since these times of the day provide some of the most dynamic listening conditions. Some club publications have even carried scholarly journal-length articles on this subject. Other folks like to look into the **solar and geophysical conditions** that affect radio frequency propagation. When you learn more about this area of interest, you will discover that we are not totally slaves to the whims of the propagation gods. Rather, we can plan our attacks on the bands to coincide with periods of favorable propagation activity and take full advantage of all it has to offer. Study in this area is not as difficult as it first sounds. We will discuss this topic at length in the shortwave section ahead.

Antennas

An extremely popular area for expertise and knowledge is **antenna design**. Wire is cheap, and even poorly designed antennas work to some extent. You can learn from mistakes without too much expense. Most anyone at any skill level can experiment with antenna configurations and designs. Most any external antenna will demonstrate some level of improved performance over the loopstick that's inside most mediumwave receivers. So go ahead and have fun. Many antenna designs have proven successful in AM monitoring. Most can be found in the various publications made available by the two mediumwave clubs we discussed earlier. You can start by constructing some of the more basic designs and then go where your imagination and knowledge take you. Maybe you will come up with a better antenna! For example, the classic AM antenna designs call for wood support structures that can be heavy and complex. You might experiment with plastic supports or mounts made of PVC pipe. There is always a new and better way to get the job done and you might just be the person to do it. Antenna design and engineering books can be tracked down to whet your understanding of this aspect of the hobby.

You will learn how to build more complex antennas for much higher performance than simple air core loops. Antennas known as **loop sense cardioid arrays** combine loop and vertical structures to result in a highly directional antenna for great DX performance. Some people experiment with **multiple long wire antennas** that are manipulated by "phasing," to increase performance and directivity. These antennas get incredible DX results.

AM antennas do not necessarily have to be mounted outside the house. This alleviates the requisite safety constraints of external antennas. High performance antennas can be used without disturbing the aesthetics of house and neighborhood.

Radio modifications

Others pursue different paths. Remember that kid in the auto shop class who always fiddled with his car to get better performance? People like that are still called Hot Rodders. The radio monitoring hobby has its own version of **hot rodding**. These are folks dedicated to improving the performance of their receivers. Few receivers are specifically designed for AM band high-performance, so many people tweak and redesign their radios. Earlier I mentioned a receiver produced by Radio Shack, the "Realistic" Long Distance AM Radio (TRF). During its heyday and years afterward, many hobbyists (including your intrepid author) spent untold hours taking this receiver circuit to its outer limits. By my count more than 30 major modifications for this receiver were printed in the various club publications. Similar attention and behavior has been bestowed upon other receivers before and since. It is simply part of the curious nature that drives radio monitors. It is possible you will catch the radio hot rodding bug too. Take the time to learn how to safely handle electronic components through books and the tutelage of an experienced hardware hacker. You can create amazing performance improvements for very little money and a little effort. You may ask yourself, "how can a humble hobbyist improve upon a receiver designed by professional engineers?" The answer is simpler than you may think. Commercial designs answer to several masters. In addition to producing good design it may be necessary to have this design perform over a wide range of uses. A radio monitor may, for instance, modify a receiver to narrow the bandwidth to improve selectivity of an audio signal that might not be as comfortable to listen to for music and local signals. But it will allow the hobbyist to copy a station he or she might not hear otherwise. The tradeoff may be worthwhile but not commercially "marketable." Then we have the "Bean Counters." Shortcuts in mass production save a few cents. A dedicated hardware hacker can restore those pennies in the right spots and produce amazing performance improvements. A tweak of the receiver's adjustment and alignment beyond the specifications set at the factory can yield significant improvement. I never had a mass-produced receiver that couldn't be improved this way. The hobbyist has more time (and interest) than the assembly lines.

Budgeting

Almost as much fun as hardware hacking is trying to figure out ways to keep the expenses of the hobby at a minimum. Sometimes trying to do things better will show you a way to do things cheaper. I know of one well-respected radio monitor with all the verifiable successes any listener could want, who operates with a hobby budget of $25 per month. At least he has since about 1985. Before that he kept to a $10 per month budget! There are lots of problems you can throw money at, but the radio hobby also offers opportunities to find ways around the finances. Mostly this can be done by working to get the most out your existing equipment. It can be tied in with getting to know your way around the used equipment world. If you can live without the latest bells and whistles, you can often find older receivers that will perform just fine for far less money. Tied in with the hardware hacking ethic is the idea that building can sometimes save money over buying. Everyone will find some aspects of this hobby that can awaken the cheapskate we all carry deep inside.

Friendships

The camaraderie and friendship of fellow radio enthusiasts is a heart-warming earmark of the hobby. If you participate in the organizational aspects of radio monitoring, you will learn that many clubs hold periodic conventions or get-togethers. Some nationally-organized groups have smaller regional or state systems that allow for sharing the monitoring hobby with other like-minded folks. Both of the national mediumwave clubs mentioned earlier hold annual conventions in various parts of the country. This is a grand opportunity to learn more about the AM-broadcast monitoring art. Most club conventions organize learning sessions and informal practical monitoring opportunities. Such gatherings teach beginning monitors how to network with others. Mediumwavers commonly operate in tip networks; essentially phone trees where hobbyists call one another when they catch something really hot on the band. This sharing of information in real time can help any monitor add a few rare signals to his or her log. Over the years, I've had the pleasure of developing many long-term friendships as the result of actively participating in the social functions held by the various radio clubs with which I've been involved.

AM band expansion

And then there is the future. The Standard AM Broadcast Band has been extended to 1700 kHz, which represents 10 new frequencies for dozens of new stations from around the country. As this portion of the band grows more active and becomes crowded, signals will disappear under one another. The prospects of this band expansion afford opportunity for signal catches in states not possible since before World War II.

The only limits in radio are those imposed by physics. Other limitations can be overcome with thought, initiative, and tenacity.

19

Are we having fun yet?

You now have just enough knowledge to make you dangerous. Fortunately, in the radio monitoring hobby, that's also just enough knowledge to have a lot of fun. You should now be able to travel up and down the AM broadcast band discovering things you never knew existed. It doesn't matter if you plan to devote your time solely to mediumwave listening or plan to use this knowledge as a springboard to other monitoring activities. This is a good place to put the book down for a while and try to apply what you've read so far. Go play with your radio! I mean that most sincerely. You need to maintain a sense of play about your radio monitoring activities or eventually your listening will become just another task to be done. This is a hobby, not a job. Go have some fun! When you get back we'll take a look at another form of radio monitoring that can take you to places far away and exotic. Shortwave radio.

Here are
some of the
tools of our trade

20

Shortwave monitoring

The shortwave listening aspect of the radio monitoring hobby is literally as big as the world! No matter what your interests, you will find some aspect of shortwave monitoring that will bring you enhanced enjoyment. Music, sports, languages, world affairs, politics, comedy, drama, art, people, pets, and places are just some of the common subjects you will discover as you enter the shortwave monitoring world.

Language myths

Let's begin by breaking down a long-held myth about shortwave radio. Contrary to popular belief, you do not need to have a knowledge of foreign languages to enjoy shortwave monitoring. Remember, many countries are seeking out a listening audience from North America as well as in other parts of the world. These countries broadcast their programming in many different languages to attract a worldwide audience. All major international broadcasters produce at least a portion of their programs in English. So you do not need to be fluent in Spanish to hear and enjoy Spanish National Radio or German to understand Deutsche Welle. Incidentally, if you are multilingual or have a strong desire to master another language, shortwave radio is a great resource for language study. Some stations even offer language courses in their native tongues and dialects. But it is very unlikely that you will ever find a time that there are not at least 20 or more stations broadcasting in English during your listening sessions.

World music

Even if you are uncomfortable with languages other than English, you can still enjoy music from around the world. Every country and culture has its unique music, both traditional and modern. You can spend countless hours listening to music that is quite different from what you normally hear on your morning drive to school or work. In addition to culturally unique music, you will find worldwide interpretations of whatever your favorite music may be. You can hear traditional American jazz broadcast by Russian performers over The Voice of Russia and country-western music played by London-based performers over the BBC. You can also enjoy classical music broadcast from orchestras in many lands. Even rock and roll takes on new shades as it comes to you out of Asia, Africa, or Australia. Music remains a truly universal language that can be enjoyed in many forms throughout the shortwave world.

World news

Many people get involved in shortwave listening because they enjoy following world news and current events. For them, hearing about a coal miners strike in France on the evening news on television just won't do. Shortwave news devotees will seek out the signals from Radio France International to get up-to-the-minute information. Once again, you must remember that many countries broadcast a portion of their programming in English. Almost 50 countries broadcast news programs in English to North America. This should satisfy the most news-hungry individual. In addition to straight news, these stations often offer in-depth commentary to help explain their version of the situation to the world. This is especially useful in those times of world conflict and turmoil. The shortwave listener can draw his or her own conclusions from broadcasts of the nations involved in the conflict and the surrounding countries. If you use your listening skills to conduct just such an examination of a world situation, you will be surprised how little information you actually receive from the more traditional evening television news.

The staff at Radio Nederland plan an international broadcast

It is important to remember why countries set up expensive, powerful shortwave stations to broadcast around the world. For the most part, it is a desire to let the world know about their country, its people, its unique history, and its culture. So you shouldn't be surprised to discover that Radio Japan broadcasts programming about Japanese culture. For many shortwave listeners, tuning in these cultural information broadcasts is a form of world travel.

Large broadcasters see themselves as truly international. These stations, such as The British Broadcasting Corporation (BBC), make special efforts to provide programming of interest to their listening audience in various parts of the world in addition to their programming about the United Kingdom. Listening to these stations makes one feel like a member of the world community.

> During Operation Desert Storm, interest in listening to shortwave broadcasts took a significant upsurge. People wanted more information than they were receiving from the evening news. After the cessation of hostilities, many of these casual listeners became radio monitors.

Sports

While some folks spend all of their time watching baseball and football on television, more and more sports have taken on an international scope. Soccer, bicycle racing, tennis, motorsports, ice hockey, and basketball all have international followings and are often given in-depth coverage by shortwave broadcasters, especially if the event is being held in their country or if their country's team is participating. If you follow any international sport, shortwave listening will give you up-to-the-minute, play-by-play coverage on a par with what traditional "American" sports get on television.

> I work with a woman who is a native of England. I once amazed her by casually quoting the Manchester United rugby scores. I showed her how to use a shortwave receiver to keep track of happenings in England. She became a listener. After a visit home, she rewarded me with an official Manchester United scarf.

Foreign domestic programming

While many shortwave signals are intended for international reception, it must be remembered that quite a few countries use shortwave as an internal, domestic service. Where we have AM/FM radios in our homes and cars, many countries depend on shortwave radio for similar news, information, and entertainment. Some people are attracted to shortwave monitoring for the opportunity to eavesdrop on this domestic programming as a way of getting to know the world. This aspect of the hobby often requires an ability to recognize, but not necessarily be fluent in, various languages. This is very challenging but not really beyond the ability of the average listener who chooses to devote a portion of his or her listening to the study of domestic broadcasts. We will talk about the advantages of language recognition later in this section of the book.

Collecting stations and countries

But one of the most common interests shared by shortwave enthusiasts the world over is the desire to "collect" new stations and countries. This collection process can take various forms. Many hobbyists choose to keep a log book to catalog the stations they have heard. More and more people take advantage of inexpensive tape recorders to capture signals from around the world. This way you can then share what you have heard with others.

If you become interested in the collection aspect of the shortwave radio hobby, you may be interested to know that many shortwave stations welcome your letters. In return for these letters reporting what you have heard, stations often send out letters or cards that serve as formal verification of reception. These verifications, known as QSLs, are highly prized among some shortwave hobbyists. In Chapter 56 of the book we will talk about how to go about sending verifications and receiving confirmations.

Without a doubt, as you grow in your understanding and appreciation of the shortwave monitoring hobby, your world view will be greatly enhanced. This is the greatest gift that monitoring shortwave has to offer.

A SMALL SAMPLING OF
SOME INTERNATIONAL BROADCASTERS
Frequencies are in kilohertz (kHz)

Russia: Voice of Russia

4940	4975	6010	6065	6130	7115	7160	7185
7280	7300	7305	7315	9505	9515	9530	9535
9570	9580	9610	9625	9640	9730	9750	9755
9765	9815	9820	9880	9895	11630	11665	11675
11690	11695	11705	11710	11730	11745	11750	11760
11765	11785	11790	11800	11805	11810	11820	11835
11840	11870	11905	11915	11925	11940	11950	11960
11970	11975	11980	11985	11995	12010	12045	12050
12055	12070	12280	13650	13725	13775	15110	15125
15130	15140	15150	15155	15180	15190	15220	15225
15265	15290	15320	15330	15340	15350	15365	15375
15395	15405	15410	15415	15420	15425	15440	15470
15480	15490	15500	15525	15535	15540	15550	15580
15590	15595	17560	17565	17570	17580	17590	17595
17600	17605	17635	17640	17645	17660	17665	17670
17675	17680	17685	17700	17710	17720	17730	17735
17740	17755	17760	17775	17805	17815	17835	17850
17860	17870	17875	17880	17890	17895	21450	21465
21480	21505	21545	21570	21585	21590	21615	21625
21630	21640	21670	21690	21725	21745	21785	21790
21820	21825	21830					

United Kingdom: British Broadcasting Corp. (BBC)

3255	3915	5965	5975	6005	6125	6135	6175
6180	6190	6195	7105	7110	7160	7180	7230
7280	7325	7405	9410	9515	9570	9575	9580
9590	9600	9630	9640	9740	9750	9760	9915
11695	11760	11820	11940	11945	11955	12095	15070
15150	15220	15260	15280	15310	15340	15360	15370
15380	15400	15420	15575	17640	17705	17790	17830
17840	17860	17880	21715				

United States of America: Voice of America (VOA)

3980	5995	6010	6035	6040	6060	6110	6130
6140	7105	7115	7170	7200	7205	7215	7265
7280	7325	7340	7405	7415	9455	9525	9530
9575	9590	9635	9645	9665	9700	9705	9760
9770	9775	11580	11705	11715	11720	11725	11760
11805	11870	11895	11915	11920	11950	11965	12035
12040	12080	13680	13710	15120	15160	15170	15180
15185	15205	15250	15255	15290	15305	15395	15410
15425	15445	15580	15600	17735	17740	17800	17820
21485	21550						

21

What can you hear on shortwave?

The shortwave broadcast band generally comprises all of the signals that can be heard from 2000 kHz through 30,000 kHz, from just above our AM broadcast band and including that portion of the radio frequency spectrum we call High Frequency. This is a world you cannot enter with your common household AM/FM receiver. It is a world where signals can reach your home from the farthest ends of the earth.

International broadcasting bands

The first areas of shortwave monitoring that most beginners pursue are the International Broadcasting Bands. As we mentioned before, there are powerful stations in many countries that send out their signals to be heard in other parts of the world. The

105

programming may be entertainment, news, sports, or political in nature. Regardless, the station's intent is to reach you—the international listener. Since these stations intend for their signals to reach your receiver, they are usually the easiest catches for folks just starting out in shortwave monitoring. Instead of one band of frequencies like the AM broadcast band, international shortwave broadcasters make use of 14 different bands, commonly referred to by a relative wavelength expressed in meters.

FIGURE 10
INTERNATIONAL BROADCAST BANDS

Frequency Range	Band
2300 - 2495 kHz	120 meters
3200 - 3400 kHz	90 meters
3900 - 4000 kHz	75 meters
4750 - 5060 kHz	60 meters
5900 - 6200 kHz	49 meters
7100 - 7350 kHz	41 meters
9400 - 9990 kHz	31 meters
11600 - 12100 kHz	25 meters
13570 - 13870 kHz	22 meters
15100 - 15800 kHz	19 meters
17480 - 17900 kHz	16 meters
18900 - 19020 kHz	15 meters
21450 - 21750 kHz	13 meters
25600 - 26100 kHz	11 meters

For example, the band of international broadcast frequencies that runs from 7100–7350 kHz is commonly called the **41 Meter Band**. As you encounter various other resources for frequency information, you might find the frequencies for these bands listed as slightly different. Don't let this keep you up nights. The world is slowly adjusting these band allocations to be in compliance with the 1992 World Administrative Radio Conference (WARC) band plan. Occasionally you will even find international broadcasters operating outside of these standard bands. This is all part of the fun and intrigue that surrounds shortwave monitoring.

As you begin to monitor, you will find many international broadcast stations operating on multiple frequencies and on different bands at different times of the day. Stations will even change the time of day of their broadcasts and their operating frequencies depending on the season of the year. This movement serves to take advantage of optimum propagation characteristics to the different parts of the world to which the station wants to send its programming. This means that tuning in your favorite programs will require a little more forethought than you apply to punching the buttons on your car radio on the way to work. But the extra effort on your part is well rewarded with the interesting programming of broadcasters from around the world.

Your first shortwave monitoring experiences will probably consist of tuning through these international broadcast bands to see what you can hear. But as I've said earlier, if you've hung with me this far, I know you are a curious type of person. You're probably already asking yourself, "What's going on in those spaces between the international broadcast bands?" Tuning between those international broadcast bands will reveal the signals that often make up the news stories that you hear on the domestic and shortwave broadcast bands as well as your TV.

Utility stations

The signals between the international broadcasters are generally referred to as "**utility stations.**" These are signals that are not really intended for you, but they can be great fun to listen to just the same. Much of what you will hear on these frequencies will be devoted to commercial operations. Modern multinational corporations require the world-spanning radio capabilities that shortwave provides to get their business done. Worldwide maritime and aircraft operations make use of the shortwave spectrum to communicate en route to their global destinations. You will discover military communications from all branches of the services of the United States and many other countries. A surprising amount of this communication occurs "in the clear" so you will be able to listen in on all the action. There are bands of frequencies allocated to amateur radio where "hams" from more than 300 countries attempt to communicate with one another.

Clandestine and spy stations

Then there are the signals that are notorious, sinister, and clothed in intrigue. You will hear strange stations that repeat long lists of "spy numbers." There are **clandestine stations** that broadcast revolutionary political ideals against existing government structures. You will also find **pirate broadcasters**, folks who put on programming without benefit of being recognized by any government licensing authority.

Non-voice signals

Already you have learned about enough frequencies to occupy even the most dedicated monitor for a lifetime. But there is much more. As you tune around to all the stations and signals that broadcast using voice transmission, you will notice almost as many signals that sound like chirps, beeps, buzzsaws, hums, and squawks. These noises are more than noise, they are some of the many non-voice communication modes that are also out there for the tenacious monitor to log. Some radio monitors acquire special equipment to translate these signals into something they can understand. Many of these signals can be translated with very little additional cost beyond the price of your shortwave receiver.

Up-to-date information

You are probably already wondering how you are going to sort through this myriad of signal opportunities. Don't let that worry you at this point. Later in this section you will learn where to get the most up-to-date frequency and signal information. This will be important because things often change so rapidly on the shortwave bands that almost any frequency information I could put in this book could become obsolete between the time this book went to print and the time you got your hands on it.

You're about to enter on a journey that will span the globe. In this section you will be equipped with the tools and the talents to make it possible to become a shortwave monitor. I hope you enjoy the trip.

22

Shortwave cost and budget issues

Now we have to talk money. If you are new to radio monitoring, it's unlikely that your receiver covers the shortwave bands. Shortwave monitoring requires a **general coverage receiver**; one that covers 150 kHz-30 MHz. Your search will disclose a wide range of prices.

General coverage receivers run the gamut from about $40 to $6000 and beyond. Obviously a $40 radio is going to lack some performance. But this does not mean that high performance is weighted at the high end of the price scale either. Actually, there are a number of relatively high performance portable general coverage receivers in the range of $150–$300. If even that's a bit too ritzy for you, the used market can cough up a good performer with judicious shopping. For example, the older Radio Shack Realistic DX-160, lacking the bells and whistles of ultramodern radios, will let you log your first 50-75 countries for about $50-$75. The Yaesu FRG-7 was one of the hottest shortwave receivers of the late 70's. This puppy logged well over 100 countries for many operators and it's now about $100. A buck a country isn't such a bad deal. I wish that principle applied across the board.

This Watkins-Johnson HF-1000 retails for $4000. Don't worry! You can fully enjoy the shortwave hobby without spending anywhere close to this amount.

Now that you have a clue on the basic cost of business, I can let you in on some good news. Antennas don't cost much. Practical antennas for the shortwave bands consist of as little as one long piece of wire. More complex antennas have two pieces of wire and a length of coaxial cable to reach the receiver. Add some static and surge protection, and your antenna system will hear everything there is to hear and ring in at less than the cost of movies with your significant other. You can, of course, spend more on commercially produced antennas ($75 or so), but shortwaves will respond just as well to home brew antennas, perhaps better with good design.

Lest you think that this hobby is expensive, compare it to others. Priced a set of golf clubs lately? How about a pair of in-line skates? A performance-quality bicycle? Bicycles run $40-$4000 too. The neat thing about a decent general coverage receiver is that you can easily extract ten years of use out of it, perhaps even twenty or more if you are not too enamored by the all latest techno-goodies. Think of it this way: a $300 receiver works out to $30 per year. Not too much for a hobby that can return so much pleasure and knowledge. You can't stretch a pair of in-line skates out that long!

So, as with most hobbies, you start with a long, hard look into the wallet, checking account, or that 5 gallon jug full of loose change. A good point of departure would be to spend 75% of your allowable funds on the receiver and the remaining 25% on the antenna, accessories, a hamburger, and a 6-pack of your favorite beverage.

23

Shortwave receivers

In order to fully enjoy shortwave monitoring, your single most important acquisition will be a shortwave receiver. In this chapter you will learn how to go about purchasing your receiver, where to make your purchase, and what standards of comparison are important in your shopping.

While it is possible to purchase a shortwave receiver from some large consumer electronics stores, most shortwave listeners prefer to deal with any one of several radio-communications specialty stores that deal in the mail-order sales of shortwave receivers and accessories. The reason why most people go this route is because of the enhanced technical support that such companies provide. Would you rather buy your equipment from someone with lifelong interest and experience in radio communication or from a department store salesperson who doesn't know shortwave from a permanent wave? To aid your search, a list of some popular vendors is in *Appendix 1.*

New or used?

The choice between a new or used receiver can be difficult, especially for a newbie to the hobby. It is possible to essentially double the buying power of your budget when purchasing a comparably-featured used receiver. However, the same warnings apply to the purchase of used radios that apply to used cars, with one notable exception. Used receivers tend to hold their resale value through several owners. This means that if you are someone who wants to try shortwave monitoring but are unsure of your interest, you could purchase a modestly-priced used receiver and then sell it for almost what you paid for it to either leave the hobby or to move up to a new receiver. In either case, a little research goes a long way in making a purchase that can last for many years.

Pre-owned receivers can be very economical

If you are in the market for a used receiver, examine the classified sections of your local newspaper or similar sections in many radio and electronics publications. Radio hobby magazines will also clue you in to the location of flea markets, swap meets, and amateur radio "Hamfests" that represent the best source of used equipment. As stated earlier, we will discuss these publications at some length later in Chapter 31. Also, some of the same specialty stores listed in *Appendix 1* deal in used gear as well as new equipment.

Whether you choose to purchase a new or used receiver, the best advice I can give at this point would be to obtain the catalogs from several shortwave supply houses and compare and contrast the various receivers until you find one that gives the best level of performance you can afford. Don't get too rattled when you see the wide range in prices between moderate- and high-priced equipment. While the advanced features of some higher-priced receivers are most certainly desirable, they are by no means necessary for full enjoyment of the shortwave hobby.

*Scrounging tubes to maintain older equipment is a
rewarding challenge*

If you choose the used receiver route you must add on a few additional considerations. Once again, think about the kind of things you would look for in the purchase of a used car. Does the receiver show signs of abuse or excessive wear? Do all the dials, knobs, and meters perform their tasks properly? Is the person you are making the purchase from trustworthy? What kind of guarantee is being provided? If at all possible, use the receiver for about an hour to get the feel for anything that might represent a problem due to overheating. During the hour trial, make a point to leave the receiver tuned to one frequency for about fifteen minutes and check for any drift off frequency. Ideally, you should try to get use of the receiver for a few days before making your final decision.

If you enter into the used receiver market, you will come to find that some of the equipment may contain vacuum-tube circuitry. Since vacuum tubes have been largely replaced by solid-state components such as transistors and integrated circuits, you must plan to spend some time and money seeking out replacement tubes that grow increasingly scarce. While I am a great fan of fine tube receivers, such equipment might frustrate a beginner. So move in the vacuum tube direction only if you have a high comfort level with old-time electronics.

> Restoring older vacuum tube equipment can be a labor of love. In 1983 I started a newsletter for devotees of the R-390A receiver. This eventually grew into *The Hollowstate Newsletter*. The current publisher, Ralph Sanserino, can be reached at 12072 Elk Blvd., Riverside, CA 92505-3835.

How to purchase a receiver

Okay, you have determined how much money you have to spend, so the next step is to obtain catalogs, publications, or other resources that tell you about the receivers in your price range. You may have noticed at this point that I am not mentioning the names of any particular units. With more than 50 receivers currently on the market, with more being produced each year, and with well in excess of 200 excellent used receiver possibilities, I would rather not limit your horizons. The major manufacturers of general-coverage shortwave receivers today are AOR, Drake, Grundig, Icom, Kenwood, Japan Radio, Lowe, Panasonic, Sony, Sangean, Siemens, Watkins-Johnson, and Yaesu.

For the moment, let's say you have gone through the catalogs and magazines and you have come up with two new and one used receiver as candidates in your particular price range. How do you decide which one to buy? To make the best decision you must consult the specifications of the receiver and compare the figures. This is not as difficult as it sounds because, once explained, the meaning of these various figures is fairly straightforward. You will find all new receivers have clearly published technical performance data and most equipment resource catalogs make a point of listing all the important information. As for that used receiver in your price range, you would have to look in the owner's manual or consult one of several resources that list specification data for used equipment. Yes, there have been entire books written on the specifications of used receivers. We will examine these resources in *Appendix 2*. With that said, we can take a look at the areas of comparison you will be gleaning from those specifications.

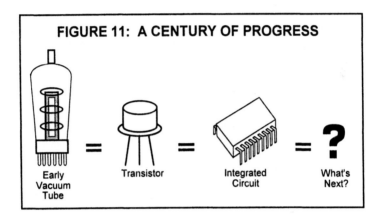

FIGURE 11: A CENTURY OF PROGRESS

Early Vacuum Tube = Transistor = Integrated Circuit = What's Next?

Coverage

Remember that we are looking for a receiver that covers from 1.6 MHz to 30 MHz. This would include the frequencies immediately above the AM broadcast band and all of the shortwave spectrum. While it is possible to purchase receivers that cover various portions of the shortwave bands, you will find that you will get the most value, use, and enjoyment out of a true general-coverage receiver. This is one that covers the entire shortwave spectrum, not just several portions of the band. Many modern receivers cover from 150 kHz through 30 MHz. These receivers tune from below the standard AM broadcast band continuously up through the entire shortwave spectrum. There are even receivers that can tune further up through the VHF and UHF spectrum but most such equipment is relatively expensive, somewhat complicated to use, and contain several design compromises, placing them out of the realm of most folks experiencing the radio monitoring hobby for the first time.

Sensitivity

Simply put, sensitivity is a receiver's ability to hear weak signals. Since signals in the shortwave spectrum vary greatly in strength, this is a very important consideration when purchasing a receiver. You will usually find the sensitivity of a receiver expressed in a quantity called **microvolts** and this is often "referenced" against 10db (**decibels**) to account for atmospheric noise and noise generated within the circuitry of the receiver itself. So you might read on the spec sheet something like "*0.5 microvolt or better for 10db signal-to-noise ratio.*" Don't let this electronic mumbo jumbo throw you! You can make an adequate comparison without mastering the theory. Some receivers will have the sensitivity figures listed for various bands or modes of operation. Again, don't let this confuse you. What you want to do is compare the figures given for the receivers that you are considering and look for the *LOWEST* number. A sensitivity figure of less than 0.5 microvolt throughout the shortwave spectrum is an excellent choice for most listeners.

> **Older general coverage receivers often did not include the standard AM broadcast band. Be careful when shopping for older receivers if you plan to do any mediumwave monitoring.**

Selectivity

Now that you have determined which receivers hear the most, you must determine which one can hear the *best*. Selectivity is the expression of how well your receiver lets you hear what you want to hear while eliminating everything else that gets in the way.

The shortwave spectrum is a very big place indeed; about 28,300 kHz wide. As you tune your receiver across the shortwave band, it listens in on little chunks of frequency. The width of the chunk that is heard at any one point in time is referred to as the tuning **bandwidth**. Many shortwave receivers can perform at various levels of selectivity (various bandwidths), depending on the type of listening you do. The bandwidths of a receiver are established by filters, filter networks, or digital signal processors that let you hear what you want while separating out nearby signals on either side of the tuned frequency. What you will want to discover is how well your receiver choices succeed at delivering the level of selectivity you need to listen to shortwave broadcast stations.

Most everyday standard AM radios found around the house have a bandwidth of 8 to 10 kHz. This is fine for general-purpose listening on a band where the stations are well separated, such as the 10 kHz spacing of the AM band. However, shortwave listening requires that you sort out stations that reside more closely together so you will want to seek out a receiver that has the ability to resolve signals in a 4 to 6 kHz bandwidth. If you intend to listen to shortwave broadcast type signals intended for your hearing by the station operator, you need look no further. However, if you plan to listen to any non-broadcast utility signals such as those heard from commercial shipping or amateur radio operators, you will want additional, more narrow, bandwidths. For this listening you will need a bandwidth of 3 kHz for single sideband (SSB) voice transmissions and at least a 0.5 kHz bandwidth for International Morse Code (CW) transmissions. Pay attention to this issue because a more narrow bandwidth is not always better. If your filtering is too narrow, the human voice becomes garbled and unintelligible. For this reason, many receivers sold exclusively for use by amateur radio operators do not make very good shortwave broadcast receivers because their filtering is not wide enough to allow a shortwave broadcast signal to be enjoyed.

Most receivers achieve their various bandwidth positions by switching in a series of filter networks. This is an area where the buyer has to beware. The receiver you are considering may have a switch on its front panel that indicates three or more bandwidth positions. However, many high-performance receivers are sold with one or two filters installed and additional filters need to be purchased if you choose to do various additional forms of listening. Be sure you clear this matter up in your comparison shopping, because additional filters can cost between $50 and $100. If you are dealing through any of the better shortwave supply houses you will also find that these vendors make available various receiver modifications that improve upon performance (also for a price). These modifications are usually very helpful but make sure that the company agrees to provide warranty coverage that does not go against the original manufacturer's guarantees.

When you see selectivity figures on a specification list there will usually be two figures, one referenced to -6 dB and one to -60 dB. Pay attention to the ratio between the two figures listed under these references. The ratio between the bandwidth measured at -6 dB and -60 dB is known as the **shape factor**. A 1:1 ratio is perfect but not possible in the real world, so you will want to look for a ratio that is close to or under 1:3 for all practical purposes.

An increasing number of receivers are appearing on the market with **digital signal processing** (DSP) features. Digital signal massaging can occur in both the radio frequency and audio frequency sections of the receiver's circuits. As this technology develops, it should be superior to current hardware filter systems.

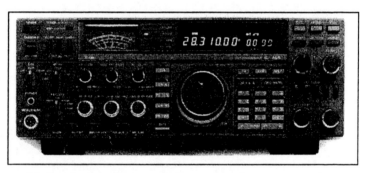

Modern ham radios usually have general coverage receivers
(Photo courtesy of Icom America, Inc.)

Receiver stages

As you read over the specification sheets for the receivers you are considering, you will probably notice references to the receiver being **single conversion, double conversion, triple conversion**, etc. These are references to the number of circuit **stages** in the receiver that process the radio signal coming in through the antenna. The technical issues surrounding receiver design can fill several books. Without getting too complicated, as a rule *the more stages the better*. Therefore, a double conversion receiver has higher performance than a single conversion rig. A triple conversion system would be better than either. You will even find receivers with **quintuple conversion**. Most moderately-priced equipment will be either double or triple conversion in design.

Stability

Stability is particularly important if you are looking at any older used receiver, especially if it is a receiver with a vacuum tube circuit. All receivers, even modern solid-state equipment, generate a certain amount of heat during normal operation. This heat can cause the tuning of the receiver to drift off frequency. Most modern solid-state receivers are designed for a drift of less than 300 Hz during the first hour of operation. This deviation is generally insignificant to most shortwave listeners. In most cases, it will not even be noticeable to all but the most exacting ears. However, significant drift in a modern receiver can be a sign of trouble in the circuitry.

Even high-quality vacuum-tube gear will exhibit a certain amount of drift, but it should resolve itself within the first 15 or 20 minutes of operation. This is what people mean when they refer to a piece of equipment being "warmed up." If the receiver drifts wildly after a reasonable warm-up period it is an indication that one or more of the tubes may need replacing; this can be a major expense in a world where replacement tubes are often scarce and costly. This is why I recommend that you get at least an hour with any piece of used vacuum-tube equipment before you hand over your money.

Dynamic Range

Dynamic range is another practical consideration. This refers to a receiver's ability to hear weak signals in the presence of strong signals generated nearby. This is very important as you begin to listen for stations that are a little more exotic and off the beaten path of the shortwave world. Poor dynamic range is usually the result of overloading by a strong station. Dynamic range is computed by several methods but most specification sheets will indicate the receiver's **blocking dynamic range**. You will want to look for a figure in the area of 100 dB.

Readout

As you begin to compare the various receivers within your price range, you will discover two types of frequency readouts. Many modern, moderately-priced receivers now have **digital readout**. This is where the frequency appears in numerical form that changes as you tune up and down the band. Lower-priced and most older receivers will have some form of **analog readout**, usually some form of pointer moving across a panel with the frequencies printed on it or a dial with the frequencies moving across a pointer. Many older analog shortwave receivers will have a main readout and a separate **bandspread** tuning readout with yet another separate control for further discriminating the frequency to which you are listening.

It is fairly simple to figure out that digital readout is generally preferable to analog readout. My only warning would be that you might find, especially in the used market, that a particular analog readout receiver might have substantially better performance specifications than its digital counterpart within a particular price range. You must remember that those digital readouts cost some portion of the overall manufacturing cost of the receiver. If you find this to be the case, please remember that everybody in the shortwave listening hobby got along just fine without digital readouts prior to around 1975. Further, digital readouts can make a listener somewhat lazy. You can get hooked into just tuning into particular familiar frequencies. This is like hitting the pretuned buttons on a car radio. With an analog readout, you have to dig around a bit. You will be surprised at how many interesting new signals are lurking around out there between your favorites.

If you do choose an analog readout receiver as one of your possible purchases, take a look at how easy it is to read the dial. Tune up a known frequency such as Time Signal Station WWV at 5 MHz or 10 MHz to see if the dial is accurate. Wiggle the dial back and forth over that known frequency to determine how much lash is in the dial. If the tuning is mushy, it will make accurate tuning difficult as you try to move between stations. Try to determine the smallest segment of frequency you can interpolate. Most high-quality analog readouts will allow you to accurately tune a frequency within 1 kHz, sufficient for all but the most fanatical listeners.

Modes

We have been using the AM/FM car radio analogy to drive home a point or two up to this point. Now it is necessary to relieve ourselves of a misnomer or two. The terms AM and FM actually do not refer to the radio itself, but rather to the **mode** of the signal being received. AM stands for **amplitude modulation** and FM refers to **frequency modulation**. These are two distinct ways of transmitting and receiving radio signals. As shortwave listeners, you will run across several other modes that will be of interest to you. Consequently, you will want to look at what modes your receiver choices are capable of intercepting.

Most shortwave broadcast stations transmit their signals using the AM mode, so you will most definitely want a receiver that offers this mode. You will find that most will, the exception being receivers designed specifically for **amateur radio** use.

Another mode that is of interest is SSB which is short for **single sideband**. This mode is used by commercial, military, and amateur radio stations where the primary focus is to transmit voice. A receiver designed to hear this mode of transmission will likely have the ability to receive both **upper sideband** and **lower sideband**, the two types of signals common to this mode. If the receiver does not have a specific mode switch that allows for this reception mode, take note to see if it has a switch marked **beat frequency oscillator** (BFO). You may find this on older used equipment. This control will also allow you to resolve SSB signals. Once again, do not let these terms throw you. I'm dishing out a lot of general information here that will take some digestion and further reading. Hang in there.

Your receiver's mode switch may include a position marked **CW** which stands for **continuous wave**. This is the mode used to receive the transmission of signals sent in the **International Morse Code**. Some people enjoy learning to receive code. If you feel that this might be of some interest to you, you will want to seek out a radio that receives this mode. Once again, the Beat Frequency Oscillator (BFO) may appear in lieu of this mode control because the BFO is also used to give CW signals their tone to allow for easy understanding.

And what about FM? Some portables will include the traditional FM radio band for local program listening. Beyond that, FM signals are seldom heard within the scope of the shortwave broadcast spectrum except at the very high end of the band above 28 MHz. Most of the FM signals in this range will be amateur radio operators and not broadcast stations. So FM is not a terribly useful mode for the shortwave listener.

Beyond the AM mode, none of these other modes may be of interest to you initially. However, in planning your purchase, additional modes can always be listed in the plus column because they increase the versatility of your receiver as you advance in the hobby.

Meters

Most of the receivers you will be considering will have at least one meter called a **signal strength meter** or **S meter**. This meter allows you to track and lock on to a station's signal. It also allows you to make note of any patterns of fade or the relative strength of interference and atmospheric noise. Some modern portables will have a row of **light emitting diodes** (LEDs) that work nearly as well as a traditional meter at the above mentioned tasks. Any receiver you may be considering that does not have a meter to measure signal strength would fall far behind its contemporaries at selection time.

Memory

Many modern digital receivers have the capacity to save a certain number of frequencies and related information in a memory circuit. This allows you to call them up at will without tuning around to find them. Some more sophisticated receivers even allow you to interface your receiver with your home computer to provide memory and control over your receiver's functions. When comparing the receivers in your price range, you will want to look for the greatest memory capacity. But remember, memory is useful but not totally necessary to the enjoyment of shortwave listening. Do not overlook other important performance features just because a particular receiver offers you a greater memory capacity. You can think of the memory issue as a good tie breaker between two otherwise equal receivers.

Timers and clocks

Many modern receivers, as well as some high-quality used equipment, will have clock/timers built into them. While by no means necessary to the enjoyment of the receiver, they can be very helpful to the listening art. If a clock is offered, it is more useful if it can be set in a **24-hour** mode as opposed to the traditional **12-hour AM/PM** mode. As we continue in our investigation of the shortwave monitoring world, you will come to appreciate the need for a 24-hour clock.

Voltages

Many beginners choose a portable receiver as their first radio. Their relatively low cost and basic controls make them an ideal choice. However, unlike the family transistor "go to the beach" AM/FM radio that can operate for six months on one nine volt battery, most shortwave portables eat up batteries at a fairly high rate. The more complex demands of the shortwave circuitry can suck a load of cells dry in a few long evenings of operation. For this reason, you will want to make sure that you choose a portable that has **dual power**, allowing you to operate off batteries or by plugging into the house current, either by its own plug or by the use of a battery eliminator.

Size And Weight

Modern microelectronics have made even the higher-priced desktop shortwave receivers only slightly larger in size and overall weight than the family VCR. If you are in the market for a used receiver, you will need to remember that some of the more venerable older receivers were quite huge by comparison. A popular military surplus shortwave receiver known as the Collins R-390A weighs in excess of 60 pounds and can strain a table top with ease, not to mention what moving it about has done to a few owners!

When taking into consideration the size of your receiver, remember that most pieces of electronic equipment expect to be operated in an environment that allows for adequate ventilation. This is of special importance with vacuum-tube equipment, which generates a significant amount of heat. Therefore, you must consider ventilation needs when figuring out where you will be locating your receiver and you should choose from a group of receivers that fit into your space consideration.

I know of one hobbyist who lived in a crime-ridden area of a major city. His entire apartment was stripped twice by burglars. They never touched his R-390A. Police said it was because it was too heavy to steal.

Antenna Issues

Most shortwave receivers require you to connect an antenna to allow successful reception of signals. Do the receivers you are looking to purchase allow for connecting your antenna in more than one fashion? For example, does the receiver have screw terminals and a coaxial cable connector? As a rule, the more ways you can go about attaching an antenna the more versatile your receiver will be when you begin to experiment with possible antenna configurations. If you are purchasing a portable, you will want to seek out one that has provisions for connecting an external antenna. This option will increase your versatility in the same way as it does for someone with a table-top receiver. Further, attaching an external antenna is the easiest way to multiply performance in most portable receivers.

Output Jacks

There will be situations when you might want to track and log a station that is particularly difficult to hear due to lack of signal strength, interference, or noise. In these situations, you will find it useful to use headphones to assist you. For this reason, you will want to make sure any receiver you consider has a **headphone jack**.

As we have already discussed in the mediumwave section of the book, tape recording is an extremely useful tool in radio monitoring. You can connect a tape recorder in various ways but the task is made easier by a dedicated **tape output jack**.

Necessary Accessories

Many general-coverage communications receivers are designed to be used in conjunction with other equipment and can be set up to perform very specific functions. This user-specific nature means that some equipment will not be self-contained. This is important to understand when examining your receiver choices. As we stated earlier, some receivers do not come with all possible bandwidth filter combinations installed. You must determine if you will need to go to any additional expense to bring the receiver up to your expectations.

Also, some communications receivers are shipped with a minimal speaker or, in some cases, no speaker at all. This is because many applications rely heavily on the use of headphones. In such a case, you may find it necessary to add a speaker to make armchair enjoyment of the shortwave world a practical pleasure.

Ergonomics

After you have covered the various mentioned criteria, you probably will have narrowed your choices down to one or two receivers. Now comes one more important part of your study. You must remember that you are going to spend many hours with this piece of equipment. How well you enjoy those hours of shortwave listening is directly related to how well you and your receiver perform together.

Your ability to use the receiver is a function of good design on the part of the manufacturer. So take a few minutes (or possibly a few days) to "grok" the receiver. Are the controls laid out in a logical manner? Is the readout and meter easy to read? Is the headphone jack on the front, side, or rear of the set? Are the controls easy to operate with either hand?

While it is of no technical significance, I would ask myself the question, "Do you want to stare at this thing for the next several years?" Ease of operation will significantly reduce any frustration you may be experiencing as you grow in knowledge and ability as a shortwave listener.

While these ergonomic issues are somewhat hard to quantify, a few minutes of thought in this area can pay dividends in your long-term enjoyment of the shortwave world.

Adding up the scores

These areas of comparison need to be weighted primarily in terms of those features that have direct effect on performance, mainly sensitivity, selectivity, and stability. These are the most important because all the other features are useless if the receiver falls down in any of these areas. If you can't hear it, you can't do anything with it! Once you have established the performance of the equipment in your price range, you can then determine how the other significant features improve the overall usefulness of the receiver. Now that you have done your homework, it is time to go shopping! Take your time, kick the tires, and drive it around the block if you can.

Final thoughts on receiver purchasing

We live in a world that supports the notion that the highest tech, state-of-the-art equipment is essential to an individual's self esteem. Almost every hobby has that handful of folks whose wallets are obviously more open than their minds. There are a few people who feel the need to look down upon anyone who does not have the latest and greatest equipment. It's hard enough for a beginner in any endeavor without needing to worry about being self conscious. My advice to the newcomer is to relax. Even if you had an unlimited budget and could purchase the most expensive and extensive receiver on the market, you would probably frustrate yourself fairly quickly. All the extended features of a high-priced receiver might not be of any use to you until you had some listening time under your belt to learn how to get the most out of those high dollar add-ons. Recent developments in shortwave circuitry and design have created some excellent portable receivers that contain almost all of the features of a high-quality tabletop model. As a beginner, you could consider purchasing a portable such as this as a way of getting your feet wet in the shortwave world.

After you grow in understanding and expertise, you can then look into a moderate-to high-priced tabletop receiver that contains the advanced features that would further enhance your listening skills. Your portable could then serve double duty as a back-up or spotting receiver. It can also be taken along on family and business trips to allow you to enjoy your hobby wherever you may go. Another alternative is to purchase a moderately-priced used receiver and gain experience in this manner, again leading up to the possibility of a new full-featured receiver after you have learned the lay of the land. The used rig could then be resold or kept as a backup.

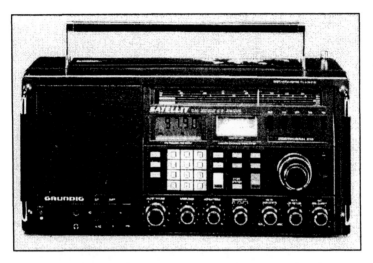

*The better quality portables can perform on a par with
tabletop radios (Photo courtesy of Grundig)*

If you already have a shortwave receiver, you might ask yourself, "When do I move up to a new rig?" After you have been involved in the listening hobby for a while, you may begin to notice that new equipment may have some performance improvements. For example, you may decide that you now need a receiver that has more modes or bandwidth filters to meet your listening needs. Perhaps there are certain special features that make a new receiver desirable. As digital readouts and memory systems became readily available in the 1980s, many people decided that these features were worth the added expense even though the receiver may not have had any significant improvement in the actual ability to hear a signal. These features were, in effect, labor-saving devices.

Once again, you have to examine your budget. Will the increased performance or additional features justify the added expense? Keep in mind that you may be able to improve the performance of your current receiver with the addition of various accessories such as antenna tuners, amplifiers, or external audio filtering systems. Maybe you can make some improvement to your antenna system that will offer increased performance for very little money. These possibilities will all be discussed in further chapters.

FIGURE 12
MODERN PORTABLES: A GOOD BUY

Like your first love, very few shortwave listeners forget their first receiver. Enjoy the search and, by all means, enjoy your radio!

24

Shortwave antennas

One of the amazing paradoxes of shortwave listening is that you can purchase a receiver for more than $4000 but still the best antenna for your application might be 50 cents worth of common household wire. Because antenna designs for shortwave monitoring are fairly inexpensive, you can have a lot of fun experimenting and playing with antennas. If your first shortwave receiver is a portable, chances are you have already logged more than a few stations with the attached whip antenna. But being the kind of curious person who is attracted to radio monitoring, you have probably already wondered about how to make use of the external antenna jack you discovered on the back or the side of your portable. If you have entered this aspect of the hobby with a desktop unit, you will need some form of antenna to get things started.

Before you attempt to connect any external antenna to any receiver, even one exclusively designed to use such an antenna, always read the receiver owner's manual carefully. The manual will discuss the proper way to make such an antenna connection. Further, it might even suggest some designs that work well for that specific unit. Some portables can experience signal overload and even damage if the external antenna system is not connected in the proper fashion. A few minutes of reading can save you quite a bit of frustration and even a few dollars.

This "antenna farm" at HCJB, Quito, Ecuador, typifies the means used by broadcasters to get their signals to your receiver

The general rule with shortwave antennas is that, in most cases, outside antennas work better than inside antennas. Also, getting the antenna as high above ground as possible usually produces some advantages. Having said this, we need to take a look at some basic safety concerns.

Safety

Antenna work almost always requires a certain amount of climbing and clambering over precarious surfaces. For example, I have to go up a ladder onto a porch roof, pull up a second ladder to get to my main roof, and then work my way over some terra-cotta to get to my main antenna connections. For these reasons, take some time to think about safety first. Use sturdy, properly constructed ladders that are long enough to reach your antenna connections without forcing you to balance on the upper three steps. Do not use or move metal ladders in the area of power lines. Your antennas should *never never never* be strung in such a way that they might ever come in contact with your electrical service. Take care with swinging your ladder around. Don't try to move a ladder that is already extended. Chances are that the ladder's point of balance will be out of your reach. This has made for some great slapstick comedies but has probably maimed or even killed a few people along the way. Take the time to reduce the ladder to its minimum size before moving it. This is yet another way to assure that the ladder will not hit a power line, crash through a window, or fall on someone's head.

Buy or borrow a tool belt. Climbing a ladder is hard enough without needing to fish through your pockets for a screwdriver or a pair of wire cutters. Plan to carry all the tools you are likely to need. This not only avoids extra trips up and down the ladder, it prevents you from trying to use the wrong tool for the job, a situation that can often damage connectors and brackets beyond repair.

You should never go climbing around or work on a ladder without someone nearby in case trouble happens. Since this person is likely to be on the ground, often right below you, they should wear a construction workers hard-hat in case you drop something their way. Nothing is more likely to turn a significant other off to your hobby than getting hit on the head with a fumbled pair of pliers.

If you need to solder any connectors, do so at ground level. You may be trying to save some steps, but you just do not have enough control of a soldering iron or torch standing on a ladder. Hanging on to a roll of solder, a soldering iron, a connector, a cable, and a ladder are tasks that would make an octopus nervous. Most connector work is usually a three-handed process to begin with, so get your hard-hatted friend on the ground to give you a hand.

One last bit of common sense. Do not string any antenna wire at a height where folks could walk into it and get "clotheslined" by it. Keep the antenna high enough so that it doesn't interfere with people or vehicles and everyone will be a lot happier in the long run.

Aluminum foil antenna

Having said all that, you probably have a few preparations to make before you get your outside antenna up. But you still want to do some listening today. There is a simple antenna design that will allow you to get started for next to no expense. It does not require that you do anything outside of your house. This may also work well for folks who have restrictions on outside antennas because of where they live.

> I used the "aluminum foil" antenna in my dorm at college and again in graduate school. I used this antenna with a third-hand Radio Shack DX-160 receiver. Dozens of stations were added to my logs with this simple station that cost, at the time, less than $75. This does NOT need to an expensive hobby!

This antenna simply consists of a length of common household aluminum foil hung with thumbtacks around one or two walls of your monitoring location. You connect your receiver to this foil antenna by way of a short length of wire attached to the foil with an alligator clip, paper clip, or thumbtack. This is not the most attractive antenna around, but I used it successfully for several years while I was living in my college dormitory. If aluminum foil hanging on the wall does not meet your aesthetic standards, you can try the same basic design with wire instead of foil. You can run the wire along the baseboards of the room so as not to draw any attention to it. There is no problem with using your imagination to develop simple indoor antenna systems. Have fun and experiment.

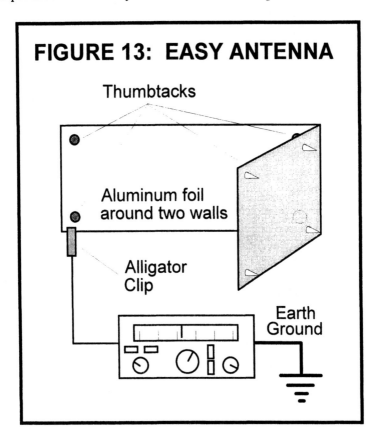

FIGURE 13: EASY ANTENNA

Longwire antennas

The first outdoor antenna most people try is the longwire antenna. As the name implies, it is a long piece of wire. The most common method of installation for this antenna is to run a length of wire, supported by insulators, from outside a window close to your monitoring post and over to a tree or other support. You will want the length of this wire to be as long as you can make it up to about 71 feet in length (71 feet represents 1/4 wavelength at 90 meters and this should work well for general listening on all shortwave broadcast frequencies). Longer than that requires some special thought because of the possibility of harmonic interference from mediumwave signals. If you can only hang about 20 or 30 feet of wire that's okay too. In most cases you will still notice that it helps out over indoor antennas.

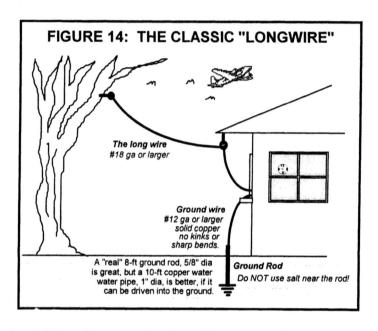

FIGURE 14: THE CLASSIC "LONGWIRE"

The long wire
#18 ga or larger

Ground wire
#12 ga or larger
solid copper
no kinks or
sharp bends.

A "real" 8-ft ground rod, 5/8" dia
is great, but a 10-ft copper water
water pipe, 1" dia, is better, if it
can be driven into the ground.

Ground Rod
Do NOT use salt near the rod!

Assembling the parts of a longwire antenna is straightforward. First you need the wire. It makes no difference if the wire is covered with insulation or not. Radio signals will get into the wire just fine. You need wire that is strong enough to withstand wind and weather without breaking. For runs up to about 50 feet or so, 14 ga or better is okay, but I'd move up to 12 or even 10 ga for longer runs. Insulated wire has the advantage of not being as subject to the corrosive effects of the outside world. Copper wire is preferred. Stranded copper has the advantage of not being as subject to kinking and breaking. Solid copper wire is probably okay for short runs but is likely to suffer from stretching and flexing that can cause it to break. One of the many corollaries to "Murphy's Law" is that your antenna wire will only break on the coldest, wettest, windiest day of the year. One common form of wire used by monitoring hobbyists with great success is called "copperweld." This is copper wire with a steel core for strength and support. It is probably the most common wire sold specifically for antenna applications by electronics supply and radio stores. If you live in an area near a rural or agricultural center, check out the local farm supply stores for copper clad "electric fence" wire. This wire tends to be strong and inexpensive. It comes in big rolls so you can do a lot of antenna experimenting.

> **I bought a ¹/₂-mile roll of electric fence wire in 1980 for $25. I still haven't used it up even though I've constructed many antennas over the years.**

Insulators are just that; simple devices that "insulate" the antenna wire from its supports. You place them between the ends of your antenna and the support structures for the wire. They can be purchased at electronic supply and radio stores. You can "roll your own" insulators out of short pieces of PVC pipe, the plastic household plumbing pipe. Insulators can be tied off by rope or wire.

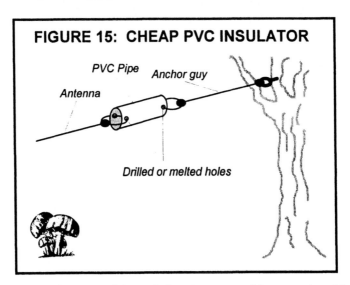

FIGURE 15: CHEAP PVC INSULATOR

PVC Pipe   Anchor guy

Antenna

Drilled or melted holes

Use rope that can withstand the elements without rotting. Nylon, polyester, or dacron rope can be found at hardware stores. Don't worry about locating hard-to-find items. A list of suppliers is in *Appendix 1*. Inventiveness and cheapskatery are acceptable practices for the monitoring hobbyist so long as you don't skimp on safety.

In the longwire application, the antenna and its lead-in to your receiver are essentially a single long piece of wire, the entire length of which serves to accept signals from all those stations to which you're trying to listen.

At this point you are probably asking yourself how you're going to safely get your antenna lead-in wires into the house to your receiver. Since you're just starting out at this game you may be reluctant to go drilling holes in window frames or through walls.

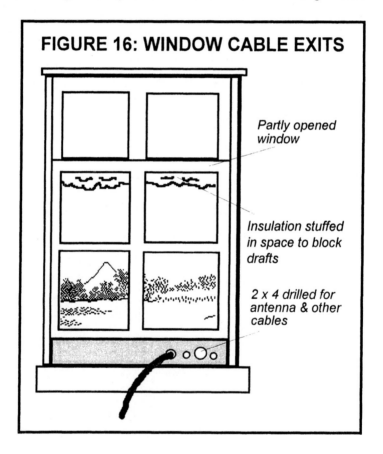

FIGURE 16: WINDOW CABLE EXITS

Partly opened window

Insulation stuffed in space to block drafts

2 x 4 drilled for antenna & other cables

A common 2" x 4" board can be cut to fit a window. Drill holes to allow wires to pass. This is a good temporary setup until you decide on a permanent scheme for your monitoring post. Some people replace a pane of glass with Plexiglas or acrylic. This lets you make holes and still keep the window fully closed.

> If you decide to run your wires through the wall,
> be sure to check for studs, power lines, and
> water pipes that may be in the wall. Drilling into
> these things can be an expensive proposition.

Dipole antennas

While the longwire antenna will work fine for most listening, you may at some point want to experiment with antennas designed for specific frequencies. Antennas can be "tuned" to do their best job at any point along the radio frequency spectrum. Remember how we discussed the direct relationship between frequency and wavelength?

Well, suppose you want an antenna that works best on the 41-meter broadcast band. This band, for our purposes, is 7100–7350 kHz; also expressed as 7.1–7.35 MHz. A resonant half-wavelength at any frequency (in feet) can be found by dividing 468 by the frequency in MHz. The center of the band of choice is 7.225 MHz so we divide 468 by 7.225; and the handy pocket calculator says this is 64.7 feet. So for our purposes a 65-ft wire is resonant near the center of the 41-meter broadcast band. *I promise that this simple formula (468 divided by the frequency in MHz = 1/2 wavelength in feet) is the only math you will be made to swallow in the shortwave antenna section of this book.*

Now scare up 65 feet of wire and cut it exactly in the middle, for the two 1/4 wavelength wires needed to construct a **dipole antenna**.

A dipole is fed at the center with an insulator separating the two $^1/_4$ wavelength pieces of wire. In this case each leg of the dipole will be 32$^1/_2$ feet. Cut the wire long enough to allow it to be secure it at the center insulator and at *both* end insulators and still leave exactly 32$^1/_2$ feet free from insulator to insulator in each section.

The dipole is then connected or "fed" at the center insulator. The most common method of connecting to a dipole is by **coaxial cable** For receive-only purposes, any radio-frequency-rated coaxial cable that you can purchase at electronics supply houses will do just fine.

Common styles are labeled RG-8U or RG-58U. RG-8U is heavier, thicker, and therefore stronger in applications where this is of concern. Coaxial cable consists of a center conductor surrounded by a plastic or foam insulator; this is then wrapped in a braided "shield" wire that is finally covered with an outer insulating material to protect the whole package. In constructing the dipole antenna, the insulation layers are stripped away from one end of the "coax" to allow you to connect the center conductor to one leg of the dipole and the braided shield to the other leg.

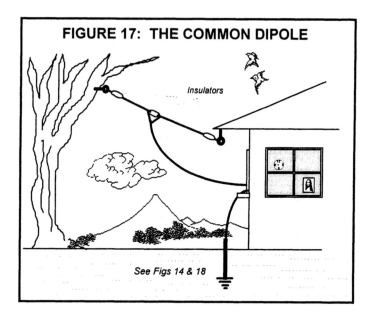

FIGURE 17: THE COMMON DIPOLE

Insulators

See Figs 14 & 18

This connection must be made so that the wires do not touch each other. The whole connection should then be wrapped in electrical tape or covered with silicone sealant to make it weatherproof. If water gets into the exposed end of the coax, it will eventually degrade. The other end of the coax is connected to the receiver by whatever style connector your manual recommends, usually a PL-259 plug. If you are not experienced with soldering, you may want to purchase coaxial cable with this connector pre-installed. If your receiver has screw or plug-in type antenna connectors, just strip the cable as you did at the antenna end and connect the center conductor to the point marked *antenna* and the braided shield to the point marked *antenna ground*. When in doubt, check your receiver's Owner's Manual.

Obviously a dipole cut for the center frequency of a band will work best at the center and less effectively toward the far ends of the band. Still, a dipole is an improvement over a simple longwire in most cases. As your understanding of the shortwave spectrum grows, you will probably want better antennas for the ranges of frequencies you are aiming for. A dipole design can do the trick.

MFJ Enterprises "GSRV" amateur antenna shows the type of construction that can be applied to shortwave antennas (Photo courtesy of MFJ Enterprises, Inc.)

Several companies produce antennas known as multiband-dipoles. These are optimized for most of the major shortwave broadcast bands. If you're not interested in rolling your own antenna, you might want to see if they suit your needs.

Active antennas

If your site limits opportunity to hang outside antennas and you don't want to string aluminum foil or wire around the inside, you have other options. Several companies make devices called **active antennas**. These are usually short whip-type antennas similar to those found on portable receivers. These are much different because the whip is connected to a preamplifier circuit that tunes and amplies incoming signals. An active antenna connects to the receiver by a short length of coaxial cable. Active antennas can offer a lot of performance in a small and unobtrusive package. There are limitations, however. First these units do cost money (usually between $60 and $100). Also, they can overload the receiver from strong local signals such as nearby mediumwave broadcast stations. Still, if you have no other antenna options, an active antenna unit can provide very high performance.

Ground connections

Some receiver and antenna applications call for an external ground connection. This is often in addition to the ground connection afforded by the standard three-pronged plug and socket in your house. A ground system is strongly recommended to drain off antenna static and to serve as minimal protection from electrical storms. Let's get one thing absolutely clear before we embark on this path. *NOTHING WITHIN YOUR POWER CAN SAFELY DISSIPATE A DIRECT LIGHTNING STRIKE!!!* Devices sold as "lightning arrestors" can only serve to dissipate the much lower-level transient voltages that appear during a lightning storm. They are very successful at doing this and I strongly urge use of such devices with external antenna systems. They can be purchased at most electronics supply stores or from the suppliers recommended in *Appendix 1*. Read and follow the installation instructions to the letter. The surest way to protect equipment from damage during a thunderstorm is to completely disconnect it and put all outside antenna and ground connections out of the house. An extra measure of precaution includes disconnecting the receiver from the AC line in your house.

Construct an outside ground with an 8-ft copper sheathed steel rod, 5/8" dia, sold through electrical supply stores. To improve ground conductivity, use two or more 8-ft ground rods, spaced 2-3 ft apart. Connect the rods to each other with short straight lengths of #8 gauge wire clamped to the top of each rod. You can achieve better results with less expensive 10-ft thick-walled 1" copper water pipe at your local plumbing supplies store. To drive the pipe into the ground, flatten and shape one end into a point. If your soil is too hard, instead of flattening the end, hook a garden hose to the top end and use water pressure to "drill" the pipe into the ground. Common stainless steel hose clamps are excellent for connecting the ground lead to a ground stake. Ground conductivity can be further improved with an 8-ft post hole filled with bentonite, a common clay, available from chemical supply firms in 100-lb sacks.

If the area of your ground rod is not regularly moistened by rain or lawn watering, pour water around the rod every week or so.

The lead-in to the ground connection on your receiver should be at least #8 gauge wire. The wire run should be as short and straight as possible. If you have to make bends, make them slow and wide, not at sharp angles. After the True Earth Ground is connected, periodically check all connections for corrosion and loosening.

So what if your site won't permit an outside ground rod? Make your ground connection to the household copper _COLD_ water pipe. Use hose clamps or a ground clamp (available at most TV and radio supply stores). Clean the pipe of corrosion. Ensure the pipe is copper all the way out into the ground. New houses and apartments may use PVC and plastic fittings around water meters. You can bridge non-conductive unions with an #8-ga wire clamped across the gaps. Use #8 gauge wire because it is a size or two larger than the other wires commonly used in a household. It provides a path of least resistance to ground. Electricity is basically lazy. If it has the choice of traveling down a great big #8 gauge wire instead of squeezing through a more common #12 gauge house wire, electricity is expected to go for the bigger wire.

Just as I explained in the mediumwave section of the book, antenna experimentation is relatively inexpensive and can be a great deal of fun. Just be sure to abide by all the safety considerations mentioned. If you have any further concerns you should also contact your local government offices to check out your building, fire, and electrical codes. When in doubt, check it out.

I know of an amateur operator whose equipment was destroyed when a lightning strike traveled up his ground system. Lightning is nothing to fool with!

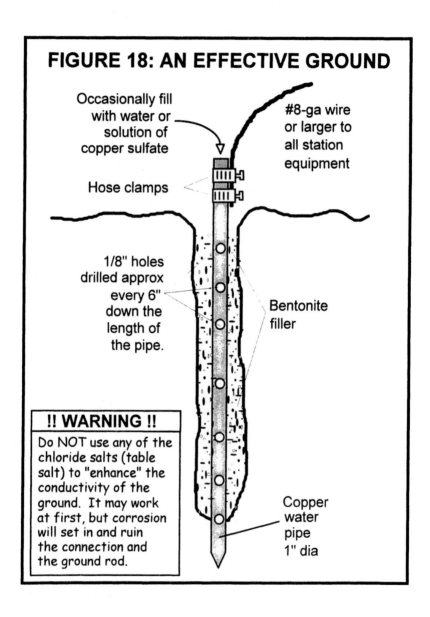

FIGURE 18: AN EFFECTIVE GROUND

Occasionally fill with water or solution of copper sulfate

#8-ga wire or larger to all station equipment

Hose clamps

1/8" holes drilled approx every 6" down the length of the pipe.

Bentonite filler

!! WARNING !!

Do NOT use any of the chloride salts (table salt) to "enhance" the conductivity of the ground. It may work at first, but corrosion will set in and ruin the connection and the ground rod.

Copper water pipe 1" dia

25

Shortwave accessories

If you think of radio in its most basic and non-technical terms, invisible signals in the air come into your receiver from the antenna. "Magic" happens inside the receiver that turns those invisible signals into sounds that can be heard from a speaker or headphones.

Accessories in the shortwave hobby mostly consist of devices that are installed somewhere along the path from antenna to speaker that serve to improve the processing of the received signal. Let's trace the path from antenna to speaker and examine some of the accessories that can be inserted along the way.

Between the antenna and receiver can be connected a device called an **antenna tuner**. This is a simple circuit that serves to match the antenna and lead-in cable to the receiver's impedance at its antenna terminals (usually 50Ω). This provides maximum efficiency in transferring the signal from the antenna to the receiver.

Another device that can be inserted at this point is a **preamplifier**. Older receivers had a noticeable drop in sensitivity toward the upper end (30 MHz) of the shortwave spectrum.

Preamplifiers boost weak signals but are tricky. You need to know when to switch it in and out of the signal path to prevent strong signal overload. Preamplifiers also boost noise that comes in with the signal. For this reason, the path between the antenna and the preamp should be as short as possible.

FIGURE 19: THE MAGIC OF RADIO

Signals arrive at the antenna.....

Magic happens....

Sound comes out...

A more sophisticated device that can be plugged into the path before the receiver is a **preselector**; essentially a tuned circuit that passes the desired band of signals while rejecting signals from adjacent bands (**harmonics**). Preselectors come with and without amplification. Even the most expensive receivers can benefit from a well-designed preselector.

If your receiver suffers from excessive overload from strong signals, another device that can be inserted in front of the receiver is an **attenuator**; essentially the opposite of an amplifier. It serves to reduce signal strength to make the desired signal come through without overloading the receiver's front end circuitry.

As you experiment with antennas, you may have more than one or two hanging at any point in time. An **antenna switch** allows you to move conveniently between two or more antennas to enhance your listening. There are switches on the market that allow you to move between multiple antennas and receivers. This affords flexibility.

As you look through shortwave radio accessory catalogs, you will find that some of the above-named devices may come packaged in a single unit. For example, you may find an antenna tuner, preamplifier, and attenuator all in one case.

There are also accessories to install inside your receiver. Some receivers are designed for user-installation of additional **filters** to improve selectivity of certain types of signals. Check your owner's manual to determine what can be done with your rig.

Several shortwave suppliers provide additional signal enhancement and filtering modifications as aftermarket add-ons. Consider these quite carefully because they often mean changing a receiver to improve some signal processing while losing others. For example, aftermarket filter modifications might increase selectivity to allow you to hear some signals at the expense of wider audio that might be desirable for stronger, easy to listen to, broadcasters.

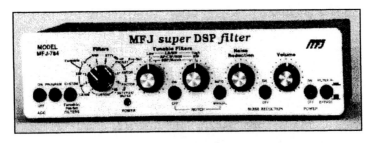

Digital signal processing clarifies noisy signals
(Photo courtesy of MFJ Enterprises, Inc.)

The audio signal out of the receiver affords another opportunity for signal processing and improvement. Circuits that shape, filter, and amplify the audio signal can improve "hearability" of more difficult signals. Recent developments in **digital signal processing** (DSP) have made audio-filtering units available to the hobbyist that in the past could only be found in professional audio-recording studios. DSP removes unwanted signals, usually noise. DSP can be used directly in the audio line or off-line with a tape recording of the incoming signal. The off-line option lets you listen to the signal over and over, making adjustments that can enhance intelligibility.

A side effect to the curiosity that makes up a radio monitor's persona is a compulsion to fiddle with gadgetry. As a beginner, you need to "attenuate" this desire a bit. Spend your time listening to signals "in the raw" for a while. You really need to get a good handle on what you are able to hear and how it all sounds before you can make effective use of signal-processing accessories. After you have had some quality time at the dials, you should be able to make reasonable decisions about what type of accessories you will add to your receiving system. You won't need or want all of the goodies I have mentioned.

Your system may look impressive with all those goodies, but there is a point of diminishing returns that can degrade the effectiveness of the station. The biggest mistake is to have your accessories on the wrong settings, causing you to miss out on what you want to hear in the first place. Don't complicate your life any more than you have to at this point in your monitoring experience. Having said this, if you are starting out with a relatively inexpensive receiver and you want to get more use out of it before you invest in a more expensive unit, then a well-chosen preselector on the front and a high-quality audio filter on the back can help a great deal.

FIGURE 20: YOUR MEGA-STATION?

Don't forget the accessory suggestions made in the mediumwave section of this book. A good **tape recorder** and monitoring quality **headphones** should still be your first accessories of choice. These two additions to your listening post will always yield the most bang for the buck.

26

Shortwave propagation

Shortwave listening does take a little more thought and effort than tuning in to your favorite talk show host while traveling in your car. This is because of the effects of propagation as they apply to the shortwave spectrum. To someone with no knowledge of shortwave, it all must seem like magic. Somebody way off in Freedonia sends out a signal that somehow gets all the way to your little box on your table top back in New Jersey. Furthermore, how come that guy in Freedonia can't just set his transmitter up on the same frequencies as your local AM station? And while we are at it, how come you can only hear Radio Free Freedonia during certain hours, on certain days, in certain seasons, even though the guy transmits 24 hours a day, 365 days a year? The answer can be summed up in one word... **propagation**! Propagation is the science (and witchcraft) of how radio waves travel between two points, often in spite of the world around them.

As we discussed in the mediumwave section of the book, the reason your local AM "All Elvis all the time" station comes in loud and clear is because you are receiving its signals via **groundwave**.

Essentially, this means that the station sends its signals out toward the horizon. Your happy home is in the path of these signals as they march toward the horizon. The signal just sort of bowls your receiver over. No muss, no fuss, no bother.

Back to Radio Freedonia. Freedonia (an imaginary country) is way over on the other side of the planet from you. You are no longer directly in the radio wave's march toward the horizon. Since it is essential to the continued political stability of Freedonia that its broadcast be heard in the US of A, Radio Free Freedonia broadcasts in the shortwave frequency range. This allows for **skywave propagation**. Freedonia's signals travel past its horizon line, bounce off the **ionosphere**, and head back down toward earth and your receiver over on the other side of the globe. Remember the pool "bank shot" analogy from the earlier section? Sometimes the signal even bounces back to earth and heads back up to bounce off the ionosphere again or makes multiple "bank shots" off of the atmosphere. This puts the signal even further away from its transmitter location.

As we said earlier, the formal name for this bouncing is **refraction**, but you will hear people call it other things such as **skip** or **path** or plain old **bounce**. See why things can get complicated?

To use the pool-table analogy again, the angle at which the radio signal hits the ionosphere directly affects where the signal will come down and be most clearly heard. So you think this would make it really easy for the stations. All they would have to do is set up antennas that would give them just the right "bank shot" to get the signal where they want it to go. To a certain degree, this is done. This is one of the big differences between mediumwave and shortwave broadcasting systems. Mediumwave station antenna systems are often designed to minimize the effects of skywave propagation to avoid interference with stations in other parts of the region. Shortwave broadcast stations design their antenna systems to take advantage of the effects of refraction to get their signal to their desired overseas audiences. However, the ionosphere changes the rules of the game by changing the height at which the signal will refract back down to earth. Imagine lining up a perfect pool shot only to have someone move the rail back a foot right after you hit the cue ball. Consequently, the science of predicting how the ionosphere is going to react becomes important to both the transmitter and receiver in the world of shortwave monitoring.

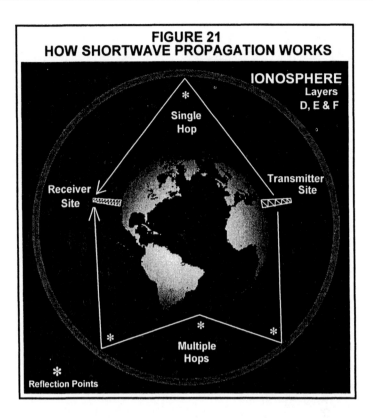

FIGURE 21
HOW SHORTWAVE PROPAGATION WORKS

Let's review what we learned in science classes and had refreshed here in the mediumwave section. This time we will dig deeper into the science. Relax, I promised to keep this stuff painless, remember?

Ionospheric regions

The area that ranges between 60 and 1200 kilometers (km) above the earth is generally considered the ionosphere. The gases between these altitudes become "ionized" by the **ultraviolet radiation** from the sun. The more radiation the more ionization occurs.

If the ionosphere is very dense, too much of the transmitted signal will be absorbed and not enough signal will get back to earth to be heard acceptably. On the other hand, if the ionization is not dense enough, the transmitted signal will not refract as desired, instead it heads off into space.

With almost a century of both professional and amateur propagation prediction under our belts, certain discoveries have allowed radio people to make good predictions about when and where to listen to hear just about anything they want to hear. The shortwave monitor's interest includes the **D**, **E**, and **F** regions.

The D Layer

The **D** layer is usually responsible for allowing your favorite shortwave broadcasters to reach your listening post each evening. As we discussed earlier, this region shows up between 50 and 100 km above the earth. Its ionization is very low at night but it becomes very densely ionized during daylight. So dense that it absorbs any signals below 7000 kHz, effectively blocking most long-distance shortwave communication. Remember how I said that events such as solar flares can even go so far as to wipe out long distance communications completely? In essence, the **D** layer is like a gate. When night falls between you and the station you are seeking, the gate is open because the absorption effects of this layer are at their minimum between these two points. The best time to listen for your favorite shortwave broadcaster is during the hours when both you and that station are in relative darkness. If you want to conduct a simple experiment, shine a flashlight on a world globe and you can get an idea of how this might work.

The E Layer

The **E** layer runs between 100 and 160 km above the earth and is responsible for most shortwave propagation. What makes this layer interesting is that it becomes effectively weaker at night. Practically you can think of it rising up to meet the **F** region. So if the **E** layer refracts radio signals at 100 km high during daylight, and this layer rises as the night goes on to about 160 km, the angle at which a signal will hit the refractive layer will change greatly. Too steep an angle and the signal travels up and is absorbed or maybe even punches through into space. The lower the angle of refraction, the greater distance the signal can travel. This layer's actual height changes with the season and with solar activity. This is why many shortwave stations make frequency changes, usually on the first Sundays of March, May, September, and November.

The F Layer

As we said in the mediumwave section, the **F** layer is out there in the sky between 160 and 320 km high, and represents the last layer of the ionosphere off which shortwave signals can normally refract. Remember, too, that this layer splits into the **F1** and **F2** layers during daylight hours.

After local sunset, the **D** layer no longer blocks long-range communication. The "gate" has been opened to the rest of the world that is in darkness. The **E** layer rises to meet the **F1** and **F2** layers that have combined into a single **F** layer around 250 km above the earth. This overall process is predictable enough to allow stations to plan frequency and antenna patterns to create maximum effective communication over desired distances. As a matter of fact, those shortwave broadcasters are depending on these phenomena to get their signal around the world to you.

Propagation charts

If you are anything like I was when I first started to monitor shortwave, you probably don't feel like spending many hours researching propagation patterns when that time could be better spent twisting the dials and hearing stations. To be honest, the real advantages of in-depth propagation analysis will only bear fruit as you move into more specialized listening habits. As a beginner and even an intermediate level shortwave monitor, you can get everything you need to know about propagation from the various radio monitoring publications. Both commercial and club magazines provide propagation information that is usually enough to get your mind right about how the effects of propagation are going to affect your monitoring habits.

The propagation charts that these publications provide usually show you the predicted **maximum usable frequency** (MUF) and the **lowest usable frequency** (LUF) for various times and frequencies. What this means is that, given all the conditions affecting all of the layers of the ionosphere, at any particular time, any frequency above a certain level (MUF) will pass through the refractive layers into space. Also, any frequency below a certain level (LUF) will be absorbed by the ionosphere and will not return to earth. Between these two figures is a window of frequencies you can use to hunt stations in a given area. As a rule, the closer a station transmits to the MUF the better its received signal strength will be. This is because the signal is less likely to be affected by the absorptive properties of the **D** and **E** layers.

We will talk more about computers and shortwave down the line, but I just want to mention one matter at this point. Inexpensive (even free) computer programs exist to help you figure out the MUF and LUF at any time and frequency. All you need to do is plug in a little information you will learn about later, when we talk about time signal stations such as WWV, and you will be able to use the same predictive information that commercial broadcasters use to make their plans to get their signal to your receiver.

Isn't there more to it? You can bet your paycheck on that, my friend! We haven't even taken a look at tracking sunspot cycles and half a dozen other factors that can affect your listening. But don't get too bent out of shape. You have the rest of your life to doodle around with propagation. There are dozens of books to absorb and theories with which to experiment. Relax! At this point in your monitoring career there are more than enough signals to catch without needing to resort to the finer points of propagation. And by the way, don't be too surprised if you hear something even when the charts say you can't. Propagation is far from an exact science. That is all part of the adventure of shortwave monitoring.

27

Shortwave
modes

One of the facts you must contend with in the world of shortwave monitoring is that there are many forms a signal can take. Let's take a look at some of these modes of transmission you are likely to experience when you first start spinning the shortwave dial.

Back in the CB boom years of the early seventies, some people used to enjoy making horses' patooties of themselves by keying down their microphones without saying anything. This crude form of jamming was known as "throwing a carrier." The **carrier** is just what it's called. It is the signal that serves to "carry" the information to your receiver. Just like those red-hot days of the CB craze, a steady, unbroken carrier is fairly useless all by itself. The carrier signal becomes the radio signal that we all know and love only when it is modified or manipulated in some way. **Modulation** is the process of manipulating the carrier signal in some way to allow it to convey information. This modulated signal is then **demodulated** by the circuits in your receiver, allowing you to hear the latest tiddly wink scores on Radio Freedonia's weekly sports program.

Clear as mud? I guess it's time to break out the surfing analogy.

Think of the ocean. If you are a life-long, landlocked dude, go to your video store and rent *The Endless Summer, Beach Blanket Bingo*, or *Surf Nazi's Must Die*. You'll get the idea. An unmodulated carrier is represented by no waves when the surf is down

Now let's visualize that the surf is up. Waves break on the beach at various heights or speeds. The bigger waves represent modulation of the ocean/carrier. But, any surfer will tell you not all waves are created equal. The different waves/modulations react differently and dropping the analogy for a second, will require different receiving equipment (demodulation if you will) to be heard and understood.

Continuous waves

Yeah, continuous waves would be every surfer's dream! But to radio folks, continuous wave signals are best known as CW, the one practical use of an unmodulated carrier. CW is the simplest form of transmission. When someone sends a signal using Morse Code, the process of hitting the key to form the dots and dashes turns the carrier signal on and off. Think of those equal height and speed waves we first talked about. Imagine they start coming in sets of three and sets of five with a lull in between. The duration of the sets and the spaces in between are like dots and dashes.

Since an unmodulated carrier has no "sound" to it, your receiver has to jump through a few hoops to give you something to copy. Some receivers have a CW position on their mode switch. Others have a BFO switch. BFO stands for **Beat Frequency Oscillator**. In either case, the switch serves to turn the "soundless" CW signal into a recognizable series of audio tones.

Learning to copy CW signals can be fun. Many ham operators still use this mode. Pick up a code practice tape from one of the suppliers mentioned in the *Appendix 1* and join the fun. I feel like James Bond or Boris Badenov when I copy CW.

> **As of the summer of 1995, The United States Coast Guard no longer routinely monitors CW signals. The mode does remain very popular with amateurs and still has some commercial uses.**

Amplitude modulation

Not long after Marconi first tossed signals into the air using the rudimentary forms of CW transmission, the early pioneers of radio discovered that you could make voice transmissions by modulating the carrier. This was done by changing the **amplitude** or height of the carrier. Wax up your surfboards and I'll explain.

You're back on the beach. Imagine that the waves keep coming toward the beach, spaced exactly 20 feet apart and that they hit the beach every 5 seconds. The only "change" you can see occurring is that all these otherwise equal waves are of different heights. Now back to the radio. This change in **amplitude** (a fancy word for height) can be interpreted as a change in voltage at the receiver that can then be translated into an audio signal. Eureka, voices that come out of the air!

But it's not always that simple. Just like at the beach, two waves can come crashing so close to each other that they crash into one another. When this happens, neither wave is of much use to the surfer. All radio signals have a certain **bandwidth** (usually expressed in kilohertz). Regulations establish these bandwidths for transmitters to assure that signals can exist on the same band with a minimum of interference. On the receiving end, many transmitters have a bandwidth switch that allows the user to vary the amount of signal heard. This serves to block out adjacent signals. So when we think of our waves at the beach we are also interested in how wide the wave is parallel to the shoreline.

Most of the shortwave broadcast signals you hear will be in the AM mode although some experiment from time to time with the next mode we will discuss.

Single sideband

If you have been tuning across the shortwave frequencies for any length of time, you have no doubt run across some signals that sound like a duck quacking. These are **single sideband** (SSB) signals. From the earliest days of radio, folks were always trying to do two things. Cram more signals into less space and get more signal from a given amount of power. SSB does both jobs remarkably. To explain this we will need to get a bit mystical with our surfing analogy.

Imagine that you are seeing the wave pattern we dreamed up for AM. Remember, waves equally spaced but with varying heights? Now imagine that the waves are generating an equal, mirror image of themselves underneath the waves. The way you generate SSB is to start with a low-power AM signal (one with waves on the top and underneath). Circuits in the transmitter serve to remove one of the side bands (the lower wave or the upper wave) and the carrier (remove the ocean; see, I told you this was going to get weird). Now the transmitter amplifies the remaining sideband (the oceanless waves become tidal waves). This produces a signal that is about four times as efficient as a regular AM signal with the same amount of power behind it. It also occupies only half the space of a standard AM signal. Almost like getting something for nothing. Not quite. On the receiving end, all you hear is that duck sound. Unless, of course, your receiver is designed to accept this form of modulation. Remember when we talked about the Beat Frequency Oscillator (BFO) in relation to CW signals? The BFO signal serves to replace the missing sideband, giving you a normal human voice instead of duck noises. Most modern receivers have done away with the BFO in favor of a mode switch that will include CW (as previously discussed) as well as **upper sideband** (or USB, that's for the waves on top) and **lower sideband** (or LSB, for the waves beneath the sea that vanished).

Some pirate and clandestine stations use these modes because they are transmitting using converted or modified amateur radio equipment. You will find that most of the "utility" stations you hear on the bands will be using the USB mode. Single Sideband is the preferred voice mode for amateur radio operators. LSB will most likely be encountered when you tune across the 40 and 80 meter frequencies and USB will show up on the 20, 15, and 10 meter ham bands. A few hams still use AM, but they are fairly rare these days.

CW, AM, and SSB make up the signals you can listen in on with nothing more than your ears. As a beginner you are going to be able to fill several log books chasing down just these three modes. However, there are other modes that you will encounter.

Radioteletype

As you are tuning around the shortwave frequencies, you will undoubtedly encounter sounds that resemble roaring motors, chirping birds, and something best described as the "beedle-beedle" noise. These are usually manifestations of **radioteletype** (RTTY). Radio teletype is used by amateur radio operators, the military, governments, news services, and various commercial entities. The goal of this mode is to transmit *text* instead of voice and music. Within the RTTY world there are about a dozen signal modes based on mutually agreed upon standards of communication. Monitoring RTTY is a subject that can take more than a single book to get a real handle on. Many radio monitoring hobbyists become fascinated with listening in on these signals. In addition to your receiver, you require the additional services of a demodulator. This is a device that can read the RTTY signals and translate them into text, just as it would appear at the intended receiving station. There are few RTTY "Broadcasts," most of the communication is "point to point" in nature. Modern digital design has produced **demodulating** units that are relatively inexpensive. Some of these devices even make use of a personal computer to manipulate the text that arrives on these signals. As you grow in your understanding of shortwave monitoring, you may want to investigate this mode further.

Facsimile and television

Long before you installed a fax machine in your office, **facsimile** (FAX), was the preferred radio mode for moving pictures from place to place. Again, special demodulating equipment is required but these pictures are out there on the shortwave bands too if you want to track them down. Amateur radio operators also move pictures back and forth by a specialized communication technique called **slow scan television** (SSTV). This mode translates a picture into an audio signal that can be transmitted by standard single sideband. These signals sound like rapidly chirping birds.

Even more exotic modes

There are at least another half dozen more exotic modes that you will encounter as you begin your search of the shortwave bands. These include such things as **packet, spread spectrum**, and **amplitude compandored single sideband**. Some of these modes are even encrypted. Maybe as you develop in your skills, you will become one of those folks who figures out how to bring access to these more exotic modes to your fellow monitors. Part of the fun of shortwave monitoring is that there is always some new path to follow in search of more and more exotic signals.

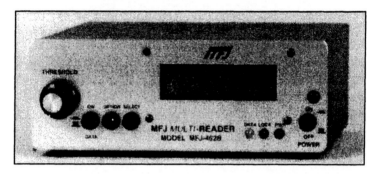

Low cost digital demodulators make easy copying of RTTY
(Photo courtesy of MFJ Enterprises, Inc.)

28

Shortwave monitoring techniques

Okay, now we get down to the meat and potatoes of the matter. Let me start out with saying something that might sound a bit strange but is true. In most cases, you will be able to log your first 25 countries in spite of yourself. How can I say this with any conviction? Quite simply, actually. To understand this you have to think back on some of the things we talked about during our discussions of antennas and propagation. *Remember, the major shortwave broadcasters out there are doing their level best to get their signal to you.* Yep, unlike long-distance mediumwave signals where the stations are actually working to *reduce* the chances of your receiving their signal, shortwave broadcasters, using sophisticated antenna systems and as many kilowatts of power as their budgets can muster, are beaming their programming right to your home. Further, most major shortwave broadcasters transmit at least some portion of their programming in English, making identification and logging as simple as if you were hearing your neighborhood mediumwave outlet. This is where you need to begin your shortwave broadcasting career. You can get your feet wet in shortwave monitoring by trying to log all those broadcasters that *intend* for their signals to get to you. In doing this, you will easily amass between 25 and 50 countries, while learning a great deal about how the shortwave spectrum behaves. Assuming you have taken my advice to start your listening with a moderately-priced portable receiver, you should be able to accomplish the goal of 50 stations without yet resorting to an external antenna. Your portable's whip antenna should do the trick for now, provided there is not any strong local interference.

Just tune around

So, very much like your first steps in mediumwave broadcast listening, just turn on your radio and tune around. Within five minutes of doing this, I will expect you to have one major question on your mind: "So how come some bands of frequencies are loaded with signals and others exhibit nothing but static?" The reason for this is that propagation characteristics for each of the shortwave bands change throughout the day and the night. You will discover bands "opening" and "closing" for your location. This is to a large degree predictable. Knowing where to listen and when to listen will be one of the first skills you will develop as you begin monitoring the shortwave bands. This, of course, is subject to the whims of propagation and the cycles of sunspots that affect the ionization of the various layers that we depend on to hear distant signals. At the risk of getting nastygrams sent to me by propagation pundits let me give you a simple rule of thumb to begin your listening. *Listen to the lower frequencies at night and the higher frequencies in the daytime.* Yeah, I can hear the experts screaming as they read this. You will discover that this rule paints a very broad brush as you grow in your understanding of radio. But that's okay. While everyone else is busy burning me in effigy you will be logging your first 25 or so countries.

41 and 49 meter bands

If you are an average person with a more or less average lifestyle you will probably be settling in for a session of shortwave listening in the evening. That works out just fine because during most evenings the 41 and 49 meter shortwave broadcast bands are open to Europe, Latin America, and some of Asia. This is where I suggest you begin your hunt. As a beginner, you will find these two ranges of frequencies to be most fruitful. You will get a good notion of who is out there trying to get their signals across to you. If you are a bit more curious, you might throw the mode switch into the Upper Sideband (USB) position and go hunting in the range of frequencies between the 41 and 49 meter bands. There you will hear some of those "point to point" utility stations in between all the "beedle-beedle" sounds of RTTY. Just do what beginners do best. Tune around and have fun. Take good notes on what you hear and where and when you hear it. If you find some programming that interests you, don't be afraid to hang around and enjoy it. You're entering a whole new world of radio broadcasting. You will begin to discover the differences and the similarities between you and folks in other parts of the world. You're becoming a shortwave monitor.

Develop your own frequency list

Later on I will give you information on how to look for certain programs or signals based upon published frequency lists. For now, get to know your way around by developing your own list of frequencies. For example, you may discover The British Broadcasting Corporation (BBC) comes in loud and clear for you every evening on 5975 kHz. Great! Keep track of this fact for future reference. The BBC is one of the most widely listened-to shortwave stations in the world. You may also discover that the Dutch World Service, Radio Nederland, comes in clearly at 6020 and 6165 kHz. Don't be too surprised about this. Many shortwave stations do what is called "parallel" broadcasting, often from more than one broadcasting site.

Relay stations

Now to bring up something that can be a bit confusing to beginners. At any given time, the programming you hear *may not be broadcast from the country producing the programming.* International broadcasters use relay stations set up around the world to get their signals to particular audience groups. For example, that signal from Radio Nederland on 6020 kHz might actually come from their transmitter site at Flevoland in The Netherlands. However, the parallel signal broadcast on 6165 kHz may come from Bonaire in The Netherlands Antilles, a group of islands in the Caribbean. That BBC signal is probably coming by way of their relay station at Antigua in the West Indies, also in the Caribbean. As a matter of fact, the majority of the BBC programming you are likely to hear comes from this station's vast organization of more than 20 overseas relay sites.

> **Don't assume that the signals you hear from Voice of America (VOA) come from the United States. VOA closed its Bethany, Ohio station, leaving domestic stations only in Greenville, North Carolina and Delano, California. However, VOA has relay stations in 19 foreign countries. Many of these can be heard in the United States. These stations can add to your country totals too. Make sure you listen long enough to determine the country you are *really* hearing.**

Is this cheating? No, what you've got here is a couple of countries that discovered they could get a better signal into the vast area of North America from the Caribbean. Should this affect your listening and enjoyment? Not one bit! The origin of these signals becomes important when you get involved in keeping track of the actual countries you hear. Thus 6020 kHz at Flevo would count as The Netherlands on your country list, but 6165 would count as The Netherlands Antilles. Two different countries for the price of one on most evenings on the east coast of North America. Keep this notion of parallel broadcasting in your mind. As you progress in the hobby, you may discover that listening in on a strong parallel station can help you to successfully identify a weaker relay station on another frequency that may represent a new country for your totals.

Bandscanning

Having used the above techniques to get oriented to shortwave broadcast station monitoring, you may want to branch out and look around for some of the other interesting signals that can be found. One technique useful to the beginner is similar to the "bandscanning" technique we discussed in the mediumwave monitoring section. Scanning the shortwave spectrum from end to end covers a lot of territory and this can lead to frustration and confusion. A better way to eat the pie is one slice at a time. Divide the shortwave frequencies into 1 MHz chunks. For example 6–7 MHz, 7–8 MHz, 8–9 MHz, etc. This gives you small, 1000 kHz portions of frequencies (remember, 1 MHz equals 1000 kHz). In this way you're doing a bandscan on a portion roughly the size of the standard AM broadcast band. This is much more manageable. Break out your log book and start hunting through the 1 MHz section of your choice. It can be any section of the band that suits your fancy, but try to pick a segment that is fairly signal-rich during the time of day you are able to listen. Go through the segment, taking notes along the way on all that you hear. Don't just track the voice or broadcast signals, pay attention to all the sounds you hear. Note the signal's mode (AM, USB, RTTY, etc.). If the signal is a voice or broadcaster, try to determine the language being used. You will find you will get the hang of what most common languages sound like even though you cannot understand what is being said. Also, try to keep track of signals that are interfering with one another: for example, a RTTY station making a broadcast station signal hard to hear. Tracking these patterns of interference may lead you to find the best time to attempt to hear a station when interference is the least. Pay special attention to the time and the frequency. It may take you up to two hours to do a thorough assessment of a busy band segment. Don't rush! The time you take will make you more savvy when it comes to digging out the more difficult signals. When you're finished, *go back and start all over*. In the world of shortwave monitoring a lot can happen in two hours. Signals come and go. Broadcasters change format, frequency, and even language. You will find that band segment analysis at different times, on different days, during different seasons, will always turn up enough changes to make the study worthwhile.

In going through band segment scans in the shortwave spectrum, you will probably discover that the time of day you normally listen, coupled with your particular receiver and antenna setup, will make for especially good listening on certain segments of the spectrum. Always play to our strengths. Make a point of becoming even more knowledgeable about those segments you hear most strongly. This serves two purposes. First, these segments will probably generate up to half of your best listening. Second, the expertise you develop can be shared with your fellow monitors through club journals and commercial magazines. Even a beginner can become a recognized expert in the shortwave monitoring community using this technique.

Spy stations

The band segment study system is the technique that many beginners can use to start logging the more exotic events that come over shortwave. In among all those broadcasters, utilities, and government stations, you will find the signals that make up the "edge" of shortwave monitoring. For example you may tune across a signal of a woman's voice repeating a long string of numbers, usually recited in discrete, identifiable number groupings. These are known in the hobby as "Spy Numbers." There is a very good chance that that is exactly what they are, too. These are coded transmissions sent for unknown purposes to unidentifiable individuals. Spy number stations are fun to log. They can appear just about anywhere but many have a time and frequency pattern that can be tracked. You'll never crack the code as it is probably a technique called a "one time pad" that would require you to have possession of a "key" document. Still, it's fun to listen in.

Pirate Stations

Also, while cruising around, particularly on weekends and holidays, you might hear broadcasts in AM or Single Sideband identifying themselves by names such as "Voice of the Unknown Kumquat" or "Radio Free Cleveland." If the music is loud and the humor is a bit off color, chances are you've run across a pirate radio station. These stations are put on the air by folks who don't feel the need to obtain proper licensing to put their signal on the air. These programs, usually transmitted by way of modified amateur radio equipment, can run from poorly produced to professional quality production. There is a whole subculture attached to pirate radio that can be as fascinating as the programming itself. If you hear one of these folks, take good notes and log the frequency. Later, in Chapter 56, we will talk about how you can confirm a pirate station. Also, pirates tend to hang out on a few frequencies, so if you hear one, hang around for a while at that spot, you may hear a few more.

"Solid Rock Radio" is one of many pirate stations on the air

Just listening

Your listening up to this point may be running along the lines of logging as many new signals and new countries as you possibly can. This is fine. Filling up the log is one of the great rushes a beginner can experience. But let me remind you of something. There is some amazing, entertaining, and informative programming going on out there on the shortwave bands. In your rush to fill all the lines in your log book, don't pass up the luxury of just plain old-fashioned listening. Programs about world news, music, and culture can be found at almost every turn on the shortwave broadcast bands.

Once you have heard an in-depth news analysis from the BBC, Swiss Radio International, Radio Nederland, or Deutsche Welle, you will never go back to the sound-bite and commercial-laden "talking head" news shows on your television. If you're a fan of any kind of music, from Punk Rock to Classical, you will find the music you like in abundance over any number of shortwave outlets. Would you be surprised if I told you that one of the best jazz programs comes out of the Voice of Russia? Or that the British Broadcasting Service has a fine Country & Western music program? You will find the music world is very small in the shortwave spectrum. These simple examples just serve as a reminder of the fun you can have listening to shortwave instead of just logging signals. Relax and enjoy yourself. Logging 100 new stations in a weekend won't bring you a raise in pay, cure the dog's fleas, or remove unsightly facial hair. This is a hobby and stopping to hear the beauty of the world along the way is all part of the fun.

During the 1995/1996 federal government "shut downs," pirate activity skyrocketed to record levels. The assumption was that, during the budget crisis, the FCC couldn't afford to listen in.

Now, at this point I'm going to take you out on a limb with me if you are the curious and tenacious person I expect you to be since you've hung in up to this point. If you have tried the experiments and techniques we discussed in the book through the mediumwave and shortwave sections then I'm ready to stick my neck out and say that you have probably come up with a few ideas of your own as to how to skin this radio monitoring cat. You have probably already amassed enough experience to try to discover new things beyond the simple suggestions we have shared up to this point. Go for it, my friend! What makes this such a great hobby is that everybody can bring some experiences to the hobby from which the rest of us can benefit. Your particular receiver and antenna, working from your unique location, against the whims of propagation as it affects your place on the earth as separated from the station you are listening to, are going to provide you with information and experiences that I'll never know about unless you tell me. After we have another talk about time, we'll discuss how you can teach me and others a thing or two through the shortwave clubs.

29

Time in the world of shortwave

In the cult movie *The Adventures of Buckaroo Banzai* (if you haven't seen it, rent it and see if you can find the shortwave radios in it), Buckaroo reminds us of that basic concept of time: "No matter where you go, there you are." Wherever you are in the world, when the sun is over your head it is noon. This worked out just fine until people started communicating over long distances. People are eating lunch in France just about the time the first light of the sun is peeking in my windows in New Jersey. Shortwave radio allows a broadcaster to sling a signal across many time zones. To reduce the confusion this could cause, common practice in the shortwave broadcasting community is to report time in terms of UTC. UTC means **universal time coordinated**. The old name for UTC was GMT, standing for **Greenwich meridian time** and was actually more explanatory of just what UTC is. If you look on a world globe or map you will see that the zero degree longitude line (longitude runs from north to south) runs through the town of Greenwich in the United Kingdom (it's a suburb of London, look really close). The time computed at this zero longitude line (sometimes referred to as the *Prime Meridian* or *Greenwich Meridian*) is the standard for UTC. All other world time zones are computed from this point by international conventions and agreements. This is one of the few matters on which just about every country can agree! For example, the time zone we refer to as Eastern Standard Time (EST) is five hours behind the zero longitude time zone. To further avoid confusion UTC is usually stated in terms of a 24-hour clock. So when it's midnight or 0000 in Greenwich, UK, it is 7:00 PM or 1900 in New Jersey. Likewise it would be 6:00 PM or 1800 Central Standard Time (CST), 5:00 PM or 1700 Mountain Standard Time, and 4:00 PM or 1600 Pacific Standard Time. Where most beginners get confused is when the United States switches over to Daylight Saving Time (DST). Our local time moves ahead an hour but UTC remains the same. That means that, during Daylight Saving Time, at midnight, 0000 UTC, EDT time would be 8:00 PM or 2000 and so on back across the country.

UTC and geophysical alerts

Want to know the current UTC? Tune to 5000, 10,000, 15,000, or 20,000 kHz, depending on the time of day and propagation. WWV, Fort Collins, Colorado will report the current **coordinated universal time** which is what they call UTC. How come nobody ever thought to call it CUT? Probably because it would make so much sense that you couldn't get all those countries to agree on it. When you send reception reports to international broadcasters, you should make your time references in UTC. This will help to avoid any confusion.

What you will hear when you listen, depending on the time and particular minute, will be a steady click like a ticking of a clock or a tone marking each second. Right before the minute you will hear a man's voice stating the time in UTC. The minute is signaled by a beep. If you hear a woman's voice instead of, or underneath, the man's voice, you are hearing WWVH, WWV's sister station in Kekaha, Hawaii. Now stop for a minute and use your new-found radio skills coupled with your curiosity. If you can hear a signal in Hawaii, what does that mean? It means a propagation opening toward the Pacific. This would indicate a good time to go hunting for Asian and Pacific rim countries, wouldn't it? WWV and WWVH have a few more things to teach you about propagation.

Geophysical alerts are broadcast on the eighteenth minute after the hour on WWV and on the forty-fifth minute after the hour on WWVH. This is where you get the information needed to make sense out of propagation for your monitoring practices. These reports include the **solar flux** and **A index** for the previous UTC day and the **K index** for Boulder, Colorado, which is updated every three hours. The bulletin includes the current state of the earth's magnetic field and predictions about conditions over the next twenty-four hour period. The folks at Fort Collins pack all this information into just 45 seconds, so it might be wise to have a tape recorder hooked up so you don't have to re-listen an hour later.

Radio Station WWV transmits time and frequency standards and signal propagation information

Don't get excited if this **A index** and **K index** stuff is gibberish to you. Scientists argue over what it all means, but a beginning radio monitor can develop a practical understanding of the meaning.

Solar flux is expressed as a number derived essentially from counting the sunspots. Remember sunspots in science class: storms on the surface of the sun. A figure of 65 or lower is typical in years of minimum solar activity. An intermediate figure is 100-200. Solar flux numbers far above or below the intermediate figure tend to indicate poorer listening conditions for shortwave monitoring.

The A index is usually the most widely discussed figure in radio monitoring circles. This is a 24-hour figure expressing the geomagnetic field of the earth. The scale runs from 0 to 400+ but rarely sees the high side of 100. Without delving too deeply into the subject, an A index of less than 10 can be considered ideal for shortwave monitoring. The lower the figure the less the signals are absorbed by the earth's geomagnetic field. Under good conditions, signals travel farther and better.

The K index can be thought of as a more up to date A index. It is updated every three hours. It is also computed with slightly different mathematics that take into account more subtle changes in the earth's magnetic field. This figure is useful for making your own predictions about how conditions are going to be over the coming 24 hours. If the K index figure is floating around 3, things should be great for shortwave monitoring.

Many other factors figure into propagation monitoring, but for the beginner, charting the A and K indexes provided by WWV and WWVH will give you a good handle on the basics. While you're checking the time, you may as well get some useful information that is often used with propagation software on personal computers to make predictions. We will talk more about this in Chapter 33.

Timekeeping

On the practical side, you will probably want to have a clock around your monitoring post that keeps UTC time. Such clocks used to be harder to track down than they are today. Many common digital clocks have switches that allow them to be set to either 12-or 24-hour format. Simply set the clock to the 24-hour format and set its face time to the signals from WWV and you're all set. If you have an "old fashioned" analog clock hanging around you can simply add another ring of numbers inside of the existing twelve to make it account for the 24-hour format. The only disadvantage to this is you may forget which ring of numbers applies at the time you are listening. Many modern shortwave receivers come with a clock circuit. These are almost always adjustable to the 24-hour format.

Get in the habit of keeping your log book time in UTC. This will help you keep track of things in the same way as do most of the stations you are hearing and as do all of your colleagues in the shortwave monitoring world.

30

Shortwave clubs and organizations

Just as with the world of mediumwave monitoring, shortwave monitoring is enhanced by networking with other monitors through clubs. Most shortwave clubs' primary activities center around the production and distribution of the club's bulletin or journal. Through such publications you will be able to share information about what your are hearing and what you're using to listen. This information goes a long way toward getting your overall signal totals up. A club affiliation will get you listening smart by tapping into the collective experience of the membership. In addition to their publications, most of the major clubs hold either regional or national meetings and conventions. You can learn more about the monitoring hobby in one convention weekend than I could show you in three books.

The American Shortwave Listener's Club

The American Shortwave Listener's Club (ASWLC) is oriented to shortwave monitors who live in the western United States. The listening opportunities and conditions are a bit different for folks that live on the Pacific side of the country. For people who are oriented to monitoring the Pacific Rim, this club makes a great deal of sense. You can obtain more information by contacting Stewart MacKenzie, 1682 Ballad Lane, Huntington Beach, CA 92649. Sample copies of the club's bulletin cost $1.

The Association of Clandestine Enthusiasts

If you are curious about pirate broadcasters and political clandestine stations you might want to consider becoming a member of The Association of Clandestine Enthusiasts (A*C*E). This club does its level best to keep up with the rapid and often unpredictable changes in the pirate and clandestine broadcasting communities. They publish a monthly bulletin called *The Ace*, which will bring you up to speed on this fascinating aspect of shortwave monitoring. In addition to clueing you in to the active pirate frequencies, they keep monitors up to date on current "mail drop" practices for confirmation of the stations you hear. For more information about this organization, send a self addressed stamped envelope (SASE) to Kirk Baxter, POB 11201, Shawnee Mission, KS 66207-0201.

The North American Shortwave Association

The North American Shortwave Association is arguably one of the largest and most active shortwave monitoring clubs in the world. In addition to publishing a comprehensive newsletter covering all aspects of shortwave broadcast monitoring, called *The NASWA Journal*, the club organizes regional meetings throughout the United States to allow for membership interaction on the local level. This club has a great deal to offer the beginning shortwave monitor in terms of information about frequencies, confirmation, and monitoring practices. More information is available by sending an SASE to Bill Oliver, 45 Wildflower Lane, Levittown, PA 19057. A sample bulletin is available for $2.

The Ontario DX Association

Canadian shortwave monitors have a club all their own that even has a few things to offer those of us in the United States. The Ontario DX Association publishes an excellent monthly journal. They also have a phone-based DX-Change information service as well as a computer BBS system. They have club meetings in several of the Canadian Provinces. For more information, contact Harold Sellers, P.O. Box 161, Station A, Willowdale, Ontario M2N 5S8, Canada.

Choosing a club

I suggest that you contact the clubs listed above and request a sample bulletin. Usually they request a nominal fee of a dollar or two. Once you have read over the samples, you will have a good idea of which groups have something to offer you. In addition to your large national clubs, there are dozens of smaller local and regional clubs that have a great deal to interest and assist the shortwave monitor. Information on these clubs and organizations can often be found through many of the commercial publications dedicated to the radio monitoring hobby.

Club membership requires some involvement on your part. You will want to become a contributor to the club's publications. Each of the club publications will explain how to go about sending your logs and other information to share with your fellow members.

the JOURNAL

NORTH AMERICAN SHORTWAVE ASSOCIATION

VOLUME XXXVI NUMBER 2   FEBRUARY 1996

ZYK 687 - 840 KHz - MW
ZYE 956 - 6.090 0 KHz - SW - 49 m
ZYE 957 - 9.645.0 KHz - SW - 31 m
ZYE 958 - 11.925 0 KHz - SW - 25 m
ZYD 819 - 96.1 MHz - FM

P.O.BOX 372 - São Paulo - Brasil

Cat. No. 62-1084

SHORTWAVE
LISTENING GUIDE

Listen to global news in the making! Monitor transmissions from Voice of America, Radio Canada, Radio Netherlands, Radio Moscow and many others. Learn how radio waves travel. Get information on popular shortwave bands and licensing for all types of equipment.

By William Barden, Jr.

Radio Shack's stock varies from year to year,
but at any given time, they have one or more references to help the
beginner get launched into the adventure of shortwave listening.

31

Shortwave publications

The two commercial magazines mentioned in the mediumwave section of this book, *Monitoring Times* and *Popular Communications*, have even more for the shortwave monitor. Each offers multiple monthly columns and feature articles dedicated to the shortwave monitoring hobby. Each caters to novice and expert alike.

The pressing issue for neophytes in shortwave monitoring is what, where, and when to listen. Shortwave stations rapidly change frequencies and programs. In recent years, even countries change overnight. Tracking all those signals has been a Herculean effort. Two books have evolved to become almost mandatory texts for surfing the shortwaves. You can hardly do without at least one of the following, and both, if you take your monitoring half-seriously.

Both books are published annually. Larger bookstores and most of the commercial advertisers in the radio magazines carry them. As a beginner, you will probably lean toward one over the other, but only because it will be the first one you get. Sooner or later, you'll taste the exceptional value that both publications offer. If not, then a good knitting and crocheting book might be better....... ☺

Passport To World Band Radio

Passport to World Band Radio, edited by Lawrence Magne, is published and revised yearly by International Broadcasting Services, Ltd. The *New York Times* calls *Passport* the "TV Guide for world band radios." As far as this is possible, *Passport* certainly fills the bill. Given the regular changes that occur throughout the shortwave broadcast world, *Passport* is the preferred text for just trying to find out what might be interesting to hear when you're not trying to dig out the internal service of Radio Freedonia. *Passport* presents a great deal of information about stations, frequencies, and programming. It goes out of its way to make the novice listener at home with the quirks of shortwave broadcasting that more experienced hobbyists take for granted. This book also introduces the newcomer to the verification process. Station addresses and confirmation information are included in a separate section.

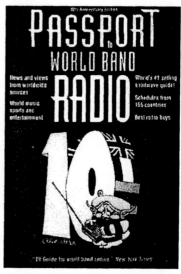

Passport's "Blue Pages" are a graphical presentation of what can be heard throughout the shortwave broadcast spectrum. The format of this section takes a little getting used to but it remains one of the fastest systems for tracking activity by frequency and time as opposed to consulting separate listings. A complete listing of "Worldwide Broadcasts in English" (by country) is also included.

Frequency and programming data are only about half the picture. Editor Larry Magne has long been known for his "Magne Tests" column in *Monitoring Times* magazine and his *RDI White Papers*. These are comprehensive laboratory analyses of the predominant shortwave receivers in the hobby. *Passport* includes Larry's investigation of the current crop of portable and desktop receivers. If you purchase a new receiver without checking *Passport* first, there could be disappointment far beyond the book's cover price.

World Radio TV Handbook

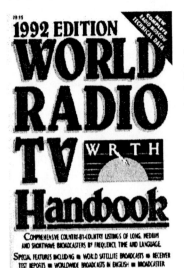

1992 EDITION
WORLD RADIO TV W.R.T.H Handbook

COMPREHENSIVE COUNTRY-BY-COUNTRY LISTINGS OF LONG, MEDIUM AND SHORTWAVE BROADCASTERS BY FREQUENCY TIME AND LANGUAGE
SPECIAL FEATURES INCLUDING ● WORLD SATELLITE BROADCASTS ● RECEIVER TEST REPORTS ● WORLDWIDE BROADCASTS IN ENGLISH ● BROADCASTER ADDRESSES AND PERSONNEL ● MAPS OF PRINCIPAL TRANSMITTER SITES

The second great resource is the *World Radio TV Handbook (WRTH)*, edited by Andrew G. Sennitt and published annually by Billboard Books. For many years the *World Radio TV Handbook* has been the single best source of information for radio hobbyists. Its comprehensive data on countries, station addresses, transmitter information, and program listings make it essential reading for hard-core monitors. The *WRTH* contains far more information than most radio hobbyists will ever need. Still, I think it's really neat to be able to look up a station's Telex number.

The hard data section of this book is divided into regions and then into countries. Each country listing gives the local time (+ or - UTC), and demographic information including primary languages. Also listed are the names of key administrative personnel at radio and TV stations as well as current addresses, phone numbers, fax, and telex lines. Stations are listed by frequency and transmitter power level. Details of operating hours and programming languages are also included.

Since confirmation is so important to many monitors, verification policy and information are included in each station listing. Station policy coupled with addresses and staff information make the *WRTH* invaluable when seeking confirmation cards and letters.

If you are really digging for those hard-to-catch signals, you will benefit from the data on transmitter locations, antenna beam headings, and target coverage areas. Another helpful feature is a listing of "Shortwave Stations of the World" by frequency. This will help the listener narrow down the possibilities when monitoring an unknown signal. Also included is a listing of international broadcasts in English by time.

Beyond the station data, the *WRTH* provides the hobbyist with other useful information including a complete list of "DX and Media" programs, "Standard Frequency and Time Signal Stations," "Broadcasting," and "Religious Organizations" lists. A good list of world-wide radio hobby clubs is also included. Also each edition is filled out with the "Listen to The World" section, a collection of articles of interest to radio hobbyists including equipment reviews.

As you grow in the hobby, you will discover dozens of books written to address every aspect of shortwave monitoring (see the selected bibliography at the back of this book). Books specializing in antenna designs abound, as do books on many of the more technical aspects of the hobby. As your interests blossom, you will pick and choose the areas of study that will allow you to advance further as a shortwave monitor.

For your immediate reference, below are regular contributing publishers on the hobby radio scene. Get copies of their catalogs about once each six months to stay up on the latest radio data:

Index Publishing Group, Inc.
3368 Governor Drive, Suite 273
San Diego, CA 92122
(619) 455-6100; fax (619) 552-9050
e-mail: ipgbooks@indexbooks.com
Web http://www.electriciti.com/~ipgbooks

CRB Research Books, Inc.
PO Box 56
Commack, NY 11725
(516) 543-9169; fax (516) 543-7486

Tiare Publications
PO Box 493T
Lake Geneva, WI 53147
(800) 420-0579
Web http://www.tiare.com

Grove Enterprises
PO Box 98
Brasstown, NC 28902-0098
(704) 837-9200; fax (704) 837-2216
E-mail: mt@grove.net
Web http://www.grove.net

32

Shortwave frequency resources

The two books discussed in the previous chapter will serve as your main resources for general shortwave broadcast frequencies. But since you are the curious individual I now know you to be, you've already noticed a few things. First, since these books are only published once a year, where do you go for information on signals that have changed in the interim? Second, broadcast stations only make up a portion of all that can be heard on shortwave. Where do you go for the hot frequencies for all those non-broadcast signals?

Well, to start with, you will be your own frequency resource. As you have been scanning the shortwave frequencies, perhaps using the 1 MHz method mentioned a while back, you have already logged a lot of information that will remain useful to you. Don't discount your own efforts. Your own log book will often be the latest information you will find. Keep referring back to your log not just as a source of pride but as a source of information.

Your next frequency resource will be the club publications, which are smaller and as such can operate with shorter deadlines than most commercial publications. Frequency changes and updates will often appear first in the club bulletins. Also, many clubs have phone trees for sharing hot tips. Some publish mid-month newsletters to give an even more narrow focus.

Association of North American Radio Clubs

A group of the larger radio clubs have joined together in an organization called the Association of North American Radio Clubs (ANARC). You can receive information about affiliated clubs by writing to 2216 Burkey Drive, Wyomissing, PA 19610-1553. Please include 2 units of appropriate return postage (2 mint first class stamps or 2 International Reply Coupons). One of their affiliated activities is the **ANARC SWL HAM NET**. Many monitors are also amateur radio operators. These folks get together on the air to swap their latest tips once a week. Even if you are not a ham, you can always listen in. This net is run on Sunday mornings at 10 AM Eastern Time on 7240 kHz in the Lower Sideband (LSB) mode. This net is best heard in the eastern half of the United States. Give it a listen and you will hear about what other folks are entering into their logs. Or better yet, think about becoming a ham yourself and getting in on the fun. We'll talk about how to do this later in Chapter 52.

ANARC also has a Web Page at http://www.anarc.org/. You might know that computers are important tools for radio hobbyists. They are also important tools for radio-monitoring clubs.

Commercial publications

The two publications we mentioned earlier, *Monitoring Times* and *Popular Communications*, both have monthly columns devoted to the latest shortwave frequencies for all facets of shortwave monitoring, including those aspects beyond the broadcast world.

DX programs

Many shortwave broadcasters have "DX" programs. These are programs devoted to the shortwave monitoring hobby. In addition to all sorts of fun and useful information, these programs have segments devoted to new signals to be heard. There are also two commercial radio monitoring programs, "**Spectrum**" and "**World of Radio.**" These programs appear on several shortwave outlets and they carry information on frequency changes as well.

Spectrum's staff during a broadcast

Telecommunications

Another growing source of up-to-date frequency information is via telecommunications. A personal computer and modem can put you in touch with many established shortwave oriented message bases. Every major commercial on-line service (America Online, Compuserve, Delphi, etc.) and many local and regional bulletin board systems (BBS) have areas devoted to giving radio monitors the latest frequency and signal information available. Fully realized, this is the area with the most potential for information exchange in the coming decade. With this in mind, take a look in the next chapter at how computers can enhance the shortwave monitoring practice in your future.

SHORTWAVE BROADCAST SAMPLER (kHz)

Iran	9022	9525	11715	11790	11910	11930	15260
	15315						
Iraq	11810	13680	15180	17940			
Israel	7465	9435	11587	11603	11605	11675	15640
	15650	17545	17575	17590			
Italy	5990	6165	7275	9575	9710	11800	15330
Ivory Coast	4940	6015	7215				
Japan	5960	5975	6035	6055	6120	6150	6165
	6185	7140	7230	9535	9580	9610	9680
	9725	9750	11705	11815	11840	11860	11880
	11885	11910	11915	11925	11930	15190	15195
	15210	15270	15355	15380	15410	17810	17845
	21610						
Jordan	9560						
Kazakhstan	3955	5035	5260	5915	5960	5970	5985
	6135	7255	9505	11825	15215	15250	15270
	15285	15315	15360	15385	17605	17715	17730
	17765	17910	21490				
Kenya	4935						
Kuwait	9840	13620					
Laos	7115						
Latvia	5935						
Lebanon	6280	6550					
Lesotho	4800						
Liberia	7225						
Lithuania	9710	17760					
Malaysia	6175	9750	15295	4950	4970	5980	7160
Malta	9765	11925					
Monaco	7385	9480					
Mongolia	7260	11850	12015				
Myanmar	5990	7185	9730				
Nepal	5005	7165					
New Zealand	6100	9675	9700	11735	11900	15120	17770
Nigeria	7255						
North Korea	6576	9325	9345	9640	9977	11335	11700
	13650	13760	13785	15130	15180	15230	15340
	17765						
Norway	5960	7165	9560	9565	9590	11745	11860
	11865	11925	15220	17860			
Pakistan	9515	11570	13590	15190	15515	15555	17530
	17705	17725	17900	21520	21730		
P. New Guinea	9675						
Philippines	11690	11995	15450	15140	17760	17840	21580
Poland	5995	6135	7270	7285	9525	11840	
Portugal	9555	9570	9600	9635	9705	9780	9815
	11840	11975	21515				

A tiny sample of signals to be heard on the shortwave broadcast bands

33

Computers and shortwave

In the mediumwave section of this book we talked about some of the basic uses of computers around the monitoring post. Word processing and data management are reasons enough to consider bringing a personal computer on line. But as they say, "You ain't seen nothin' yet!" Computers take on additional tasks in the realm of shortwave monitoring.

Finding frequencies

In the previous propagation and time sections, we noted that information was available to the monitor that would allow for some prediction of conditions and suitable frequencies. Concentrating your listening on those frequencies that are most suited to your location and time of day is the most efficient way to skin the DX cat. The information gleaned from the A and K indexes off of WWV or WWVH can be massaged through a few mathematical formulas to give you the **maximum** and **lowest usable frequencies** (MUF & LUF) for a given time of day and location. Now, these formulas would probably scare away most of us who have trouble balancing our checkbooks. That's what computers are for, my friend! There are commercial and "shareware" programs available that allow you to enter a few bits of information, such as the indexes for today, and then the programs crunch the numbers for you. Out pops the best frequencies for your monitoring session. This ability really comes into its own when you are tracking a rare or hard-to-hear signal.

Control applications

Another exciting way computers have come into their own in monitoring posts is through control applications related to the shortwave receivers themselves. Many modern desktop shortwave receivers have some form of microprocessor control within their design. This is usually dedicated to such aspects of the receiver as frequency generation and perhaps even signal processing. Such receivers often have a data transfer port on their back panel. This port allows you to connect the receiver to your personal computer's data transfer port. The personal computer then has the ability to communicate with the receiver's internal "computer." Throw in some software, usually available from the receiver manufacturer, and you have almost infinite control over your receiver. These software packages allow you to perform all of the functions of the receiver from your computer's keyboard. That's right, I know a couple of you more curious types are already considering the ramifications of remote receiver control. Yep, that's all part of the deal. In addition to simple control mimicry, most of these programs allow you to take further advantage of your personal computer's more generous memory, often increasing your receiver's effective memory by an order of ten or more. Why settle for 50 frequency memories when you could have 64,000 scanned at high speed? These programs also allow you to perform "search and store" studies similar to our previously discussed 1 MHz band sweeping. Only this time your computer automatically logs all the frequencies on which it detected signals. Even more amazing, this whole system of computer/receiver interfacing is still in its relative infancy! Over the next ten years or so we will probably see even more advances in shortwave/computer symbiosis.

Demodulator units

If you want to translate those bleeps, burps, and beedle-beedles into text you can understand, there are demodulator units that operate as stand-alone devices or in conjunction with your personal computer to turn RTTY and other non-voice signals into some great monitoring. Many utility monitors specialize in this area that was once out of reach to the hobbyist. The microprocessor revolution has made non-voice signal hunting a reasonably inexpensive proposition.

On-line applications

Shortwave hobbyists have jumped into the telecommunications revolution with both feet. Using a computer and modem connected to a telephone line, monitors of all types are traveling to both commercial and private bulletin board systems around the world to share and acquire information about stations and frequencies. The major shortwave broadcast stations are not ignorant of this fact and many of them have come "on line" with addresses on the Internet and with sites on the World Wide Web. Don't let these computer terms scare you off. You don't *need* a computer just yet. When you decide you *want* one, you will be able to get up to speed without too much trouble. Personal computers have become so pervasive, you may even have expertise in this area well in excess of your shortwave skills at this point. Either way, just relax and keep your eyes open. If you stick with the shortwave-monitoring hobby, you will most likely apply the same tenacity and desire to learn to the world of computers. You'll do just fine.

A sample scenario

Let me give you a scenario that goes on every day in many monitoring posts. First you turn on your receiver and monitor WWV and get today's A and K index information. You take this information and enter it into a propagation analysis program to get the range of usable frequencies for your evening's monitoring session. Armed with this frequency range, you turn to your computer's frequency and station database to locate those stations you are interested in trying for in that frequency range. You also use your computer to check several on-line shortwave bulletin boards to get the latest frequency information. Again, you use the computer to extract the frequency and station information you need and massage it into a format that your receiver's data port can understand. Then you instruct the computer to load these frequencies into your receiver's memory and then to scan those frequencies for incoming signals. When a signal is found the receiver sends information back to the computer that alerts you to what that signal might be, based upon the information you have already stored in your system. You can then instruct the receiver, by way of the computer, to examine those frequencies you desire. While you are listening, you may even be taking notes using your computer's word processing program in order to prepare your confirmation letter. You will probably also enter your latest catches into your computerized logging program.

Get the idea? A personal computer represents a very shortwave-friendly piece of technology. There is one downside that I must share with you. Many personal computers generate a *lot* of electronic interference that can put some noise onto your signal path. You will probably find that you will have to turn your computer off or locate it at some distance from your receiver to allow you to hear those weak signals. Ah, yes, that brings us back to remote-control technology. Who knows what your solution to these problems might be? I know you'll find one that works for you. You're a radio monitor, and you won't let a little static stand in your way for long!

34

Advancing your shortwave skills

You really don't have to rush into anything. If you've been following along to this point and already have a shortwave receiver, chances are you are happy as a clam tracking down signal-rich environments and logging till the cows come home. Relax and enjoy the world of shortwave monitoring. No pressure, my friend, this is a hobby. But we cannot neglect that natural curiosity that brought you to the radio hobby in the first place. After awhile, you will probably want to try some new things that will further your monitoring skills.

Specialization

In the midst of listening, you may be drawn to certain types of signals. Some pursue either broadcast or utility signals exclusively. For others, it's fun to run with an interest in one form or mode of monitoring and develop special expertise in other areas. Some people are drawn to signals from specific geographies. For example, the programming and monitoring conditions in South America can warrant special attention. Just devote a bit of extra study and listening in a particular direction and you can quickly become an expert. This expertise can be helpful to your fellow aficionados. For instance, I have a friend who tries to stay current on radio activities in the Balkan countries. If I hear something from that region that doesn't quite make sense, I'll give him a call or drop him some e-mail for the latest information based upon his expertise.

Specialization needn't be limited to geography. Some operators are known for their skills at tracking "spy number" stations. Others go for pirates or clandestine stations. Others may enjoy listening in on the many types of aviation signals that appear on the shortwaves. Some further specialize in either civilian or military aircraft. Likewise some folks are hooked on maritime signals. There are even shortwavers who follow the activities of amateur radio operators. The sky (and propagation) are the limit. Pursue long enough what you enjoy and you'll become expert in those areas. To keep your general listening skills sharp, occasionally deviate from your skill paths and just tune around other areas with which you are less familiar. This helps to keep from getting stale. And who knows, you may even get turned on to yet another area of monitoring.

Awards achievement

Another fun pursuit is awards achievement. Lots of shortwave clubs have awards programs. There are certificates for achieving certain quantities of station confirmations. For example, there are awards for the number of countries heard. Obvious achievements worthy of their own "wallpaper" might be 50, 75, 100, and 100+ countries. Other awards might be based on logging a certain number of stations in a particular region or continent. Still other awards are issued for logging and confirming certain types of broadcasters, such as religious stations. Chasing awards can be a fun and practical way of setting goals. The process of achieving an award forces a focus on listening, to account for issues such as propagation. It's a great way to get more organized in monitoring. And, of course, those awards look great hung on the walls of your monitoring post. Once you align your interests with a club or two, you might give some thought to going after some of those awards.

The major "Awards" program in North America these days is run by the North American Shortwave Association (NASWA). A booklet about their awards program is available by sending $2 to their headquarters at 45 Wildflower Road, Levittown, PA 19057.

Learning about languages

Another activity presents itself by virtue of the worldwide nature of shortwave. If you have listened for more than about five minutes, you probably noticed quite a few languages around the bands. You don't have to learn a language other than your native tongue to enjoy shortwave, but you will learn about languages if you really want to increase your loggings. As I said earlier, you can probably log your first 50 or so countries in the English language with no problem, and with a push you may even reach that magic number of 100. Meanwhile there are probably another easy 20 or so countries right under your nose if you can just figure out a few things about the Spanish language. All of Central and South America await you! As you study the loggings of your fellow hobbyists in club journals and commercial magazines, you will get a handle on the catch words and identification phrases that will help identify stations in Spanish, French, German, and other languages. Knowledge of a few simple phrases will improve the quality and quantity of your logs.

Further, when you seek confirmation from internal-service shortwave stations, your correspondence will need to be in the native tongue of the broadcaster. Don't let this worry you. Your fellow monitors have already sorted this all out. Shortwave clubs and commercial suppliers make available pre-formatted letters and confirmation cards in most of the major languages. This makes confirmation correspondence a simple task.

You might even attempt to master a new language by way of shortwave monitoring. It can be and is done. Check your library for language lesson tapes. Spend time with such tapes, supplemented by monitoring the language of your choice in real time on the shortwave bands, to heighten the learning curve. Some shortwave broadcasters even offer language courses over the air supported by text material that can be requested. This is just one more way that shortwave can broaden your horizons.

THE 7th ANNUAL
WINTER SWL FESTIVAL

★ ★ ★

March 11-13, 1994

★ ★ ★

Holiday Inn
Kulpsville, Pennsylvania

Shortwave monitors meet to share radio lore

DXpeditions

Every monitoring location has its weakness. For example, your site might not support large antennas. Perhaps your site is in a high-noise area or near a broadcast station that causes overload. Even a relatively good monitoring post can suffer from antenna directivity. What can you do about these compromises? Take your monitoring to the road! Hobbyists refer to such activities as "*Dxpeditions.*" All you need are a site that affords improvement over your normal monitoring location, receivers, and a huge roll of wire. Shortwavers are fond of *Beverage Parties*, which might sound like running off into the woods with a keg of some adult "sipping fluid." But at my kind of beverage party you can be a teetotaler and have tons of fun.

We're talking about a *"Beverage antenna,"* named after its inventor, Dr. H. H. Beverage; an extremely long, longwire antenna; usually one wavelength, but can be several wavelengths for the desired frequency. For example a one wavelength antenna for the 41-meter broadcast band would be 130-ft long. Antennas can be strung to multiples of 260, 390, 520 ft, etc. When you get into construction of "wave" antennas, you'll learn that their optimum height is between 6 and 8 feet off the ground. This lends itself well to temporary antenna structures. Maybe you know someone who lives out in the country with a few acres of spare ground.

FIGURE 22: BEVERAGE ANTENNA ANYONE?

A very long, longwire....

A Beverage antenna can bring in some incredible signals during an evening. String out a roll of light gauge magnet wire over a few trees, rooftops, or wood posts to support the drooping longwire and you're all set for the best listening around. Some radio clubs organize antenna parties like this. You can hook up with a few hobbyists in your club to set up your own.

Listening when traveling

Don't think you should only play radio when you can string a thousand feet of wire. If you are like most newbies, you're using a portable receiver as a teething ring. So take it along when you travel. You'll be surprised what a difference even a few miles can make to your monitoring opportunities. I've always thought that shortwave monitoring is a great business person's hobby. Work that involves traveling will take you to new monitoring locations. Shortwaving sure beats watching repeats on your motel television.

Transmitting

As you monitor the shortwaves, you may gravitate toward amateur radio. This worldwide fraternity of radio hobbyists will welcome your participation with open arms. If you think you have developed a respect for propagation conditions as a monitor, wait till you start putting your own signals out as a ham operator. Now you have to worry about how well signals propagate in *both* directions. But that is all part of the fun. Later, in Chapter 52, we will talk about how you can get in on the act with an amateur radio license.

Broaden your horizons

One of the things I enjoy doing when monitoring shortwave is expand my understanding of the world around me. Broadcasters provide programming that helps me understand a country's perspectives, customs, and people. Some do it overtly with propaganda. But many do it just by being on the air and allowing me to snoop on their inside. Music that you'll never hear on your AM and FM car radios is routine on shortwave. You'll learn that we probably have more in common than what separates us. Your discoveries may even apply to your daily life. In other words, shortwave monitoring is a way to have fun and to learn. If you are still in school or have family members who are, shortwave can provide some really interesting classroom projects.

35

Are we still having fun?

Shortwave monitoring won't cure male pattern baldness or burn unsightly cellulite. But it will change you in other ways. You will understand your world better than those "other" folks who depend on 30-second sound bites on the evening news for their perspective. Whatever level you start on, whatever equipment you chose as a beginner, you will find adventure that only people who share the passion for shortwave can understand. I hope you find this aspect of the radio monitoring hobby as fun and fulfilling as I have.

So once again it's time to put this book down for a while and get some monitoring under your belt. Go have some fun listening to shortwave signals. I'll be here when you get back to show you how to play radio on the high side of 30 MHz.

ADVANCED BEGINNER'S PROJECT
THE CARPET LOOP II SHORTWAVE ANTENNA

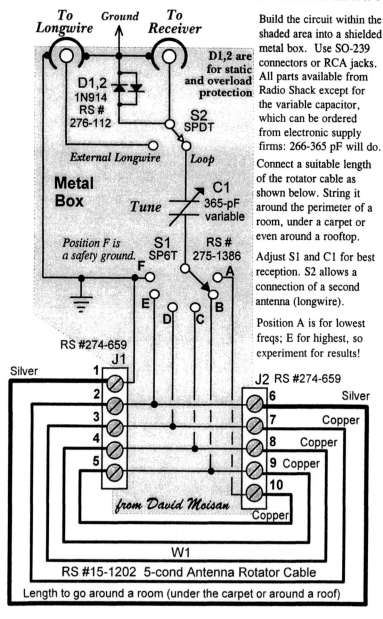

Build the circuit within the shaded area into a shielded metal box. Use SO-239 connectors or RCA jacks. All parts available from Radio Shack except for the variable capacitor, which can be ordered from electronic supply firms: 266-365 pF will do.

Connect a suitable length of the rotator cable as shown below. String it around the perimeter of a room, under a carpet or even around a rooftop.

Adjust S1 and C1 for best reception. S2 allows a connection of a second antenna (longwire).

Position A is for lowest freqs; E for highest, so experiment for results!

36

VHF/UHF monitoring

As we move up the radio frequency spectrum to the world above 30 MHz, we encounter an interesting situation. Even most folks who are not interested in the radio monitoring hobby have encountered what many people call the "Police Scanner." Actually what they know by that name is more rightly called a **scanning receiver** or more commonly a **scanner**. There are many more VHF/UHF scanning receivers sold in the United States than there are shortwave radios. Many folks own scanners but only a portion of them would identify themselves as radio monitors. Many people enjoy listening in to their local dozen or so public safety frequencies without ever exploring all that the world above 30 MHz has to offer. If you have hung with me this far, you are definitely a radio frequency explorer. You are likely to be the kind of person who will push the listening limits and seek out all there is to hear in the VHF/UHF range and beyond.

As I stated toward the beginning of the book, 30 MHz is often viewed as a sort of "line of demarcation" between two different radio worlds. It is thought of this way for several reasons. The most obvious is that the type of receiver most commonly used employs frequency scanning as the primary tuning method, as opposed to more traditional "dial" tuning (although this feature does appear on some receivers). Also, with the exception of aircraft, the majority of communications above 30 MHz occur in the Frequency Modulation (FM) mode. Another difference is in the area of radio signal propagation. In some cases, the conditions that have a negative effect on long-range communications below 30 MHz actually serve to enhance communication in the VHF/UHF range.

It's no wonder that this crossover point also represents a separation between two groups of radio hobbyists. After trying out the various aspects of radio monitoring, you may find yourself drawn to listening either above or below the 30 MHz point. There is nothing wrong with that, this is a hobby you can enjoy on your own terms. However, many folks want to have it all. They monitor on both sides of the line and in effect double their fun. These are the hobbyists who refer to themselves as "DC to Daylight" monitors.

The world of VHF/UHF monitoring has many riches to offer. Traditionally, when folks think of scanning, they relate stories about the excitement they hear when tuning in to their local police, fire, and other public safety services. But this is really only the tip of the iceberg in terms of what there is to monitor. It also misses something else that is important. Beyond the vicarious pleasures of eavesdropping on your local constabulary, VHF/UHF monitoring provides the monitor with *information*. Further, this information usually becomes available in "real time," allowing you to react to it as necessary. More than once I have been able to change my commuting route because I learned of a traffic jam or accident blocking my normal course. Even such minor bits of information can serve to make life more simple while providing you with a great deal of monitoring fun.

But the kick for most serious VHF/UHF monitors is trying to hear transmissions that other folks may not even know are there. It means, for example, moving out from your traditional local police frequencies to seek out that little-known "drug interdiction" frequency used when several departments are working together on a case. Once you catch the VHF/UHF monitoring bug, your scanning receiver will never be relegated to the same few frequencies again.

37

What you can hear on VHF/UHF

The close-in world around us does its work on VHF/UHF. Monitoring brings that world up close and personal. You will encounter the reality of this personal environment in ways never imagined through the studious use of a VHF/UHF scanner.

Law enforcement communications

As said earlier, the unenlightened think of monitoring VHF/UHF as *"police scanning."* This is only one of hundreds of sidelines in scanning. Still, for many it is an important one. It's downright *fun* to snoop on law enforcement communications, never mind much of it is shortened speech, code words, and numbers. A little attention to the flow of police talk reveals enough to understand its nature. Police departments use one or more primary frequencies for common communication, and several others for the special needs of detectives, SWAT, and traffic patrols. Most areas have at least one frequency allocated for "Mutual Aid" to allow diverse groups to coordinate on common tasks and missions.

In addition to snooping on the beat squad cars, you can eavesdrop on other law enforcement activities, such as county and state police. Prosecutors' offices, corrections facilities or jails, sheriffs, marshals, process servers, and highway patrols can be heard, if within your area of coverage. You can monitor the FBI, DEA, and other federal agencies, including the Secret Service. If the President or other government officials frolic in your area, their communications may be heard. The Environmental Protection Agency's enforcement arm can be snooped in places. In times of major disaster and emergency, these agencies work together on commonly agreed frequencies.

Private law enforcement is a hot one. Businesses employ security agencies to guard the workplace. Their signals can be very interesting. Tune to a shopping mall's security frequencies during a holiday shopping rush to turn up some exciting monitoring.

Fire fighting

Fire communication is ever a keen interest to scannists. Fire fighting, prevention, and suppression is always exciting. The larger a fire department, the more complex its radio operation. The fire radio system in a metro area might have dozens of frequencies, each with a specific purpose. Fire fighters depend on each other in larger fire situations. Mutual aid, command, and control are the key to keeping track of what's going on during fire emergencies.

County, state, and federal lands/forests are usually protected by their own fire services. These operations differ in some ways from city fire fighting. Specialized fire units might be near airports; likewise with industrial plants and factories. Fire police units are assigned their own frequencies in some areas. It's no wonder that fire fighting is one of the most recognized aspects of scanning.

Emergency medical services

The last decade or so has signaled the rise of Emergency Medical Service (EMS) units that are distinct from police and fire activities. This is an extension of what was once called the "Ambulance Squad" that may or may not have been attached to local fire services.

Ambulances and rescue vehicles, staffed by Emergency Medical Technicians and Paramedics, can be monitored as they go about the business of saving lives.

Disaster services

No one looks forward to disasters such as earthquakes, tornadoes, hurricanes, riots, and other civil disturbances. However, when these tragedies occur, the dedicated scannist can track the heart beat of the search, rescue, and recovery operation. The Red Cross, Federal Emergency Management Agency (FEMA), and amateur radio operators can be heard at these times. The Amateur Radio Emergency Service (ARES) and the Radio Amateur Civil Emergency Service (RACES) are especially apparent during crises. If an emergency involves aircraft, the Federal Aviation Agency (FAA) and the Civil Air Patrol (CAP) will be in action.

W2KUU is the ARES/RACES station in Burlington County, NJ

Military communications

A huge swath of the VHF/UHF spectrum is assigned to the military. Military installations and the National Guard offer unique scanning opportunities. Similarly, the U.S. Coast Guard operates in the maritime slice of the VHF/UHF spectrum.

Maritime communications

Speaking of the U.S. Coast Guard, if you live near any waterway large enough to have boat traffic, you will want to consider monitoring the maritime portions of the VHF/UHF spectrum for all aspects of maritime activity. These communications occur between 156.25 and 157.425 MHz and from 216 through 220 MHz. This second frequency group is the home of the recently developed Automated Maritime Telecommunications System (AMTS).

Aircraft communications

While we are in a transportation mode, we can't forget about aircraft communications. Civilian aircraft operations are in the band, 108–137 MHz with voice between 118–137 MHz. VOR and other navigation aids are on 108–118 MHz. If you are near a military installation or see military aircraft in the airspace above your monitoring post, you should listen in to 225–400 MHz where military aircraft do their business. In order to monitor aircraft communications, you will need a scanning receiver that is capable of receiving those signals in the AM mode.

Satellite signals

Also in the sky are satellites. Some can be heard with nothing more complicated than a traditional scanning receiver. Still others can be heard with just the addition of a properly designed outside antenna. Hearing signals from space brings out the excitement of radio like nothing else can. And as you start monitoring, you will discover that it's really not all that hard to do.

You don't need an antenna like this to hear satellites

Government communications

Uncle Sam is all over the VHF/UHF bands. Just about every Federal agency has a few frequencies attached to it. Entire books devoted exclusively to Federal radio signal monitoring have been published. If you want to get an idea of how your tax dollars are being spent, monitoring the government is a great hobby. Radio use is not the exclusive realm of the Feds, either. State, county, and local governments all make use of radio to conduct their day-to-day activities.

Related to local government signal monitoring are the signals from area public utility services. In addition to their everyday activities, public utilities often are called upon to respond to accidents, fires, and other problems. Many savvy scannists keep these frequencies programmed in right along with their local public safety frequencies, in order to get the whole picture of what is going on in a local emergency situation.

Business communications

In addition to all of these governmental and quasi-governmental activities, there is a whole body of frequencies devoted to businesses. Every business activity you can imagine has probably been enhanced somewhere by the addition of radio communications. For example, I once took one of my sons to a birthday party at a large commercial children's play area. Basically it was a large warehouse type of building that was filled with party rooms, video games, and a large play area for the kids to run around in and climb on. All of the internal building communications were conducted by way of handheld radio units. Businesses large and small make use of radio, and the tenacious scanner monitor can listen in on these activities with little difficulty.

Popular itinerant and low-power business frequencies include: 35.020 151.505 151.625 154.570 154.600 158.400 451.800 456.800 464.500 464.550 MHz.

Sports communications

If you are a dedicated follower of just about any professional sports activity from baseball to motorsports, you will find radio taking on more and more use since real-time communications improves performance. The world of automobile racing has probably taken this the furthest. Bringing a scanner to the race track has become almost essential for full enjoyment of the sport.

In this section we have only touched on the high points of what can be heard. As you grow in your VHF/UHF monitoring practices, you will discover that the world around you can all show up in your monitoring post by the punch of a few buttons.

38

VHF/UHF cost and budget issues

VHF/UHF monitoring can be considered as moderately priced in the world of the radio hobby. Scanning receivers can be purchased for as little as $120 new and prices can go all the way up to the fairly expensive neighborhood of $2000. The major differences from one end of the price spectrum to the other have to do with frequency storage (more memories mean more cost), and frequency management (how you manipulate all those memories). That low-end scanner may have only 10 or 20 frequency memories; this would hardly allow you to cover the activities of your local police and fire departments. At the high end you have features such as 1000 memories that can be managed and massaged by way of a computer interface. Most folks who are starting out in serious VHF/UHF monitoring activities begin with a moderately-priced handheld unit that manages about 200 frequencies. Such units sell in the neighborhood of $300. They are essentially self contained and can serve you well for years. As you move up into the range of $500 and above, you will find desktop scanning units that provide many features that are highly desirable to the most dedicated scannists. If you decide that VHF/UHF monitoring is where you want to make your hobby stand, you will probably discover the advantages of having both a desktop unit with a good external antenna system and a basic handheld unit for scanning when you are away from your main unit. Speaking of external antenna systems, these can range from inexpensive home-brew systems costing around $25 and go up to the limits of your pocketbook. The reason for this is that many scannists discover the advantage of having multiple antennas. Often at least one of these antennas is directional and mounted with a rotator for aiming purposes. Also, with VHF/UHF monitoring, there is no substitute for height. This can mean the overwhelming desire to mount your antennas on commercially-produced tower structures. But as a beginner, you can take this rabid obsession with antennas one step at a time and move only as fast as your wallet allows. You'll still have tons of fun.

39

Evaluating VHF/UHF scanning receivers

We discussed how to evaluate receivers for other areas of monitoring. The world of VHF/UHF presents significant differences that must be considered before buying a scanning receiver.

Just the term *scanner* should set your mind in a certain direction. Modern VHF/UHF monitoring is done with receivers that "*scan*" a number of programmable channels, and "*search*" across bands of frequencies. When a signal is present on one of these frequencies the scanner stops and allows you to listen to the signal's audio. When the signal stops, the scanner continues its search of the frequencies. The earliest scanning receivers had as few as four channels that could be scanned at a time and no *search* mode. A frequency was put to the scanner by a quartz crystal plugged into a socket.

Changing frequencies meant purchasing additional crystals and swapping them in and out of the sockets. Fortunately, scannists did not have to put up with this state of technology for very long. Modern digital circuit development created scanners that could store frequencies in memory chips. To change the frequencies you simply reprogrammed the memory of the scanner. This was and is usually done by way of a numerical keypad on the scanner, though computer interfaces are now available to automate this process.

> **The first scanning receivers available to the hobbyist had a few as four channels and required replacing crystals to obtain the desired frequencies.**

Memory

An important consideration for a scanner is **memory**. A scanner's memory capacity will be one of the major deciding factors when you purchase a scanner. Scanners are available with as few as ten programmable memories. However, such a unit would be *very* frustrating to a VHF/UHF monitor. Only ten memory channels would not allow any real flexibility in listening. You would spend more time entering frequencies than monitoring. Scanning receivers with 100-200 memory channels are more in line with monitoring needs. Scanner memory tends to match price. The better desktop scanners have 400-1000 channels. Modifications are possible that push the memory capacity of some scanners to over 25,000! So does this all mean that more memory is better? Well, yes and no. Higher memory capacity must be mated with other features.

Scanning rate

The scanning rate can be an important concern as you increase the number of memory channels. Using a ridiculously slow rate to show this point, let's imagine a 100 channel scanner. If the scanner operated with a scanning rate of 1 channel every second, it would take a minute and 40 seconds for the receiver to go through all the channels. Now, a lot can happen in a little over a minute and a half so you might find that you missed out on some things on other channels while the scanner was working through the programmed frequencies.

Now speed things up to a channel every half second and you're down to under a minute. Things still can slip through without a trace. Modern scanners smoke along on the order of 20-100 channels per second. Even at 10 channels per second you are through the stack in 10 seconds. You are likely to pick up transmissions you would have missed at slower speeds. Obviously, the more memory channels you are scanning the more important scan rate becomes. So when you go scanner shopping, pay attention to the number of memory channels and the scanning rate.

Programmability

Programmability is an important aspect of scanner operations. Some inexpensive scanners limit your ability to program in frequencies of your choice. Scanners such as these will only let you enter a limited number of frequencies, often from a preprogrammed, unmodifiable list. If the signal you want to hear is not in the permanent memory of this type of scanner, you will not be able to use it. So, if buying your first scanning receiver, look for a unit that has maximum flexibility in programming. You should be able to enter any frequency within the scanner's range from the keypad, in addition to being able to fill up the memories in logical groups or banks. Most scanners have a function to switch portions of these stored memories (called *banks*) in and out. These banks are usually divided into groups of 10 or 20 channels per bank. The advantage here is that you can program police frequencies in one bank, fire frequencies in the next, military in the next, etc. You can then quickly switch among the frequencies that you want to monitor.

Mode selection

It is extremely desirable to be able to adjust the *mode*. The common modes used in VHF/UHF monitoring include AM, Narrow Band FM (NFM), and Wide Band FM (WFM). High priced equipment might also offer CW and SSB modes. These are modes used mostly by amateur radio operators and only become important if you had a strong desire to monitor ham activities.

Frequency and channel control

Better scanners will "*lock out*" channels that are not needed at the moment but that you want to keep in memory. Related to this feature is another ability called "*lock-out review*" that allows you to see which channels have been locked out. Further, a good unit will have a *delay* function that can be switched on per channel, to allow the receiver to pause briefly after a signal stops, so you can have a chance of catching the reply before the scanner moves on.

Also it is useful to be able to adjust the scanner's search *step* increments. When scanners *search* a band of frequencies, there are certain conventions or spacings typically used in different parts of the VHF/UHF spectrum. For example, frequency assignments in the FM Broadcast band of 88-108 MHz are 200 kHz apart. VHF-Lo band assignments are usually 20 kHz apart. UHF assignments are typically 12.5 kHz apart, and cellular frequencies are spaced 30 kHz. Better scanners will have user-selectable step increments to cover the more common spacings. Since it is in the nature of radio monitoring to look beyond the usual, the ability to adjust these frequency steps is a decided advantage.

Still related to memory management, most scanners have one or more channels that can be designated as *priority channels*. A priority channel is periodically sampled, regardless of whatever else the scanner may be doing. This improves chances of hearing critical signals. Incidentally, a way to accomplish this in the absence of a dedicated priority channel is to simply enter your important frequency several times throughout your scanner's memory structure. It will come up more frequently this way so you will reduce the chances of missing something important.

These frequency storage and management features may seem like a lot of confusing bells and whistles when you first start out. Don't worry too much about these things right now. As you start shopping for your scanning receiver, you will discover that most of the above mentioned features are pretty much standard across the industry on all but the least expensive scanners. Just keep in mind that you want all the memory you can afford and as many ways to fiddle with it as you can get, and then you should stay on the right track.

Frequency coverage

Frequency coverage is another important aspect of choosing your scanner. Of course you want a scanner that covers as many frequencies as possible. The minimum coverage, even for low-priced units, is: VHF-Lo (30–50 MHz), VHF-Hi (150.8–174 MHz), and the traditional UHF segment (450–470 MHz). A fairly common range of frequencies for a moderately priced scanner would be from 25-1300 MHz. Most desirable are receivers that offer **continuous coverage** throughout all of these frequencies. However the TDDRA of 1994 (Telephone Dispute and Disclosure Resolution Act) made it illegal for manufacturers to market scanners that cover the **cellular mobile telephone** (CMT) segments of the 800 MHz band.

The only legal way these days to get *continuous coverage* is to buy an older model scanner that either covers this frequency range or to find a model that can be modified to include this coverage. Sacrifice of the CMT segments of the 800 MHz band is not the only gap. If you want to monitor military aviation, that large VHF/UHF band segment 225-400 MHz, you need to be sure that your chosen receiver includes this coverage. This range is often omitted in even better midrange scanners, and it cannot be *"restored"* as CMT can be on some scanners.

Do not be misled here: The TDDRA of 1994 not only forbids CMT-capable receivers, but it also forbids receivers that can be "easily modified" to receive cellular signals. In a word, if your ideal scanner was made or imported after March, 1994, it will not be feasible to modify it to receive the CMT bands.

Monitor-oriented features

Scanners that appeal to the serious monitoring hobbyist also have functions that permit *searching* through user-defined segments of the VHF/UHF spectrum in search of active frequencies that are not already entered into the scanner's memory positions. A highly desirable version of this feature is known as **search/store**. This function allows you to enter searched frequencies directly into memory for later review and use. You will find this feature on some higher-priced scanners and you may also make use of after-market add-on circuitry that allows some scanning receivers to work with a personal computer to perform this function. We will talk more about this later in Chapter 49.

> Some "high end" scanning receivers even come with a jack for video output because they cover the common television video frequencies.

Scanner performance specifications

In the shortwave section we talked a bit about specifications that make for a desirable receiver. Most of these issues apply in the world of scanners too, with a few variations as to the figures involved. Scanning receiver **sensitivity** (the ability of the receiver to hear weak signals) should be on the order of 0.5 microvolt or less and the typical selectivity (the ability of the receiver to reject interference from stations on nearby frequencies) should be 30 kHz or less. A much more critical concern is the receiver's ability to reject **intermodulation**. Intermodulation occurs when two or more strong signals work together to overload the receiver's circuitry. This is largely a product of your own environment. The best receiver in the world will have a hard time doing its job if you live next door to a powerful paging transmitter. Scanning receivers are also subject to **image frequencies**. This occurs when a strong signal mixes with the receiver's internal circuitry to produce a signal at another frequency altogether. You will want to look for a specification called **IF rejection**. The first figure listed will be the **intermediate frequency** or IF for short. The second figure listed will be the **attenuation**, the ability to reject these strong signals that cause images. Attenuation is usually expressed in terms of decibels (dB), and you will want to look for a figure of 50 dB or greater in your search for a scanning receiver.

All modern scanners depend on highly-sophisticated digital circuitry, so every receiver generates a certain amount of internal digital noise. This noise can result in what are known as "**birdies.**"

A birdie locks up the scanner as if it were a real signal. In fact, it is a "real" signal, but it originates inside the receiver. The Owner's Manual will often show a list of known birdie frequencies for a given scanner. In the case of weak birdies, you can usually adjust the **squelch** control and work around them. The squelch control cuts audio to suppress background noise when there is no signal input. If the birdie is not too strong, turning up the squelch slightly will allow the receiver to continue to scan. In the case of stronger birdies, you have to use the Lock-Out control to take them out of the action. Birdies can come from external sources, too, such as other radio equipment and accessories at your monitoring post. Usually, all you need to do to eliminate these external signals is to move the offending piece of equipment a few feet away from your scanner.

A wide variety of scanning receivers are available from AOR, Icom, Radio Shack, Uniden, and Yupiteru. You will find that these units have more similarities than differences. Careful consideration of the specs will yield the most bang for the buck.

Hot off the assembly line just before this book went to press is a new concept in receivers: *WinRadio* from Rosetta Laboratories in Australia. *WinRadio* is a PC card that plugs into your IBM/clone computer and is controlled by software. Early evaluations and reports indicate that *WinRadio* is a decent performer that offers a piece of it all at an affordable cost. See *Appendix 1* for sourcing.

WinRadio is a 500 kHz-1300 MHz scanning receiver
on a PC/AT computer card controlled by software

Radio Shack's Scanner Hall of Fame

PRO-2006 - The late classic

PRO-26

PRO-2035 & PRO-2042

PRO-2002
An early classic

PRO-43 (pre-1994)
Classic handheld

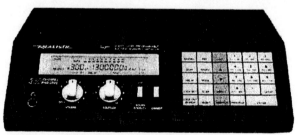

PRO-2004 - The all-time great

40

VHF/UHF antennas

Integral antennas

Just like the other forms of radio monitoring, VHF/UHF scanning performance can be enhanced with attention to the choice of antennas. Most scanning receivers come with a relatively simple antenna; usually a telescoping **whip** similar to that found on AM/FM portable radios. If the scanner is a handheld model, it will usually be equipped with a short, rubber-covered, semi-flexible antenna affectionately known as a **rubber ducky**. Now, you might think, "How can such a teeny weeny antenna do me any good when I needed to string 60 feet of wire to get any results out of my shortwave receiver?" Well, think back to our discussion of antenna length and its relationship to frequency. The higher the frequency, the shorter the antenna has to be. Remember back in the shortwave section of the book, we talked about a resonant half wavelength at 7.225 MHz being 64.7 feet? Let's play the same math game but apply it to a common area of the VHF band. If we divide 468 by 150

MHz we get a resonant half wavelength of 3.12 feet. That's a lot shorter, isn't it? Now take a look at the UHF range. If we divide 468 by 450 MHz we get a workable antenna that is only 1.04 feet long. And of course up higher where more and more radio activity is occurring, divide 468 by 900 MHz and we are down to an antenna that is a mere 6.5 inches long. That is why those antennas you see mounted on people's cars for cellular telephone operation can be so small. They are designed to resonate in the 800 MHz range.

If you were to take a common "rubber ducky" antenna apart, you would discover that it contains a piece of wire that is creatively coiled to resonate at the designed operating frequencies of the scanning receiver. If you were to stretch the wire all the way out straight, it probably would be no longer than about two feet. Since antennas can be so short at these higher frequencies, many radio monitoring hobbyists enjoy experimenting with antenna designs.

Typical VHF vertical antenna for the 2-m amateur band

There are a few tradeoffs to this short-antenna advantage. The higher you go in frequency the more critical small changes in antenna length can become. For example, we said that an antenna cut for 450 MHz would be 1.04 feet. Now apply the same formula to 455 MHz and you get an ideal length of 1.02 feet. This means that being off in your measurements by as little as a quarter of an inch might be critical to your application. So just remember the old adage to "measure twice and cut once."

The other "disadvantage" you have to work around at VHF and higher frequencies is signal losses brought about by the length of your antenna's feedline. This is always a big tradeoff at frequencies above 30 MHz. You see, in the VHF/UHF ranges, you try to improve antenna performance by getting your antenna as high up as possible (we discussed "line of sight" in the propagation sections of this book). But, as your antenna feedline gets longer, it introduces an overall greater loss in the signal getting to your receiver. These losses are usually measured in decibels (dB). Again, the higher you go in frequency, the more of a problem this becomes. This is most often combated by using the highest quality feedline you can afford in your application. We'll talk about this more shortly in this chapter.

> **When scanner monitors meet at radio conventions, they always ask for rooms on the top floor of the hotel.**

Having said all this, experimenting with VHF/UHF antennas can be fun, without stringing miles of wire around your neighborhood, as you learn how radio signals get to the receiver. If you are just starting out, you can have a great deal of fun just using the antenna that comes with your scanner. You can also do a very simple thing for a noticeable improvement in signal capture but won't cost you one additional penny. Simply take your scanning receiver to a physically higher location. In some areas, just moving from the first to the second floor gives you a handful of signals you couldn't get downstairs. When I travel, I often take a handheld scanner with me. I make a point of asking for my hotel room on the highest possible floor. Even if you don't travel much, accessible tall buildings can provide improved listening. If you think you can hear more by going up one flight of stairs, try getting up ten stories or so. What you are doing is moving the antenna to a higher location for better performance; in this case, though, the receiver stays attached to it.

External antennas

The other route many folks go as they advance in their scanning activities is to put up an external antenna, often starting out with a commercially-produced unit. There are three common designs you will discover as you look through scanner supply catalogs.

Vertical antennas

The vertical antenna is probably the most common, usually a simple vertical piece of metal, often with three or four horizontal pieces of metal around the base of the antenna. Almost all VHF/UHF antennas are vertical because most transmitted signals are in the vertical plane. Polarity is the plane of an RF wave as radiated from the transmitting antenna. A vertically polarized signal captured by a horizontal antenna will be greatly weakened. It is the electrical equivalent of the old *Three Stooges* movies where Curly walks through a 3-ft wide door with a 5-ft wide board across his chest.

If you want a common vertical antenna, read its design specifications and carefully look for two specifics. First, is the antenna designed to cover the frequency ranges you want to monitor? Many vertical antennas are optimized to give good coverage of several of the more common popular scanning frequency ranges. Again, be careful: these antennas usually lean toward the popular public safety frequencies. If you are particularly interested in, for instance, aircraft frequencies, you need to make sure that these are covered as well. Another specification to consider is any rating of **antenna gain**. This will be a figure expressed in decibels (dB), as stated earlier. Knowing this number will factor into your consideration of dealing with feedline losses down the road.

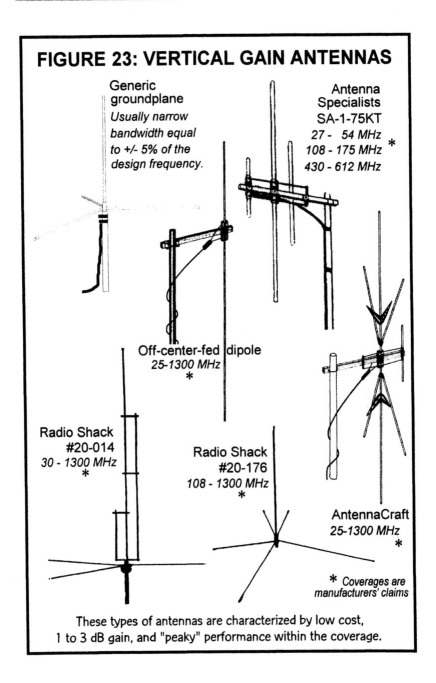

FIGURE 23: VERTICAL GAIN ANTENNAS

Generic
groundplane
*Usually narrow
bandwidth equal
to +/- 5% of the
design frequency.*

Antenna
Specialists
SA-1-75KT
*27 -   54 MHz
108 - 175 MHz *
430 - 612 MHz*

Off-center-fed dipole
25-1300 MHz
*

Radio Shack
#20-014
30 - 1300 MHz
*

Radio Shack
#20-176
108 - 1300 MHz
*

AntennaCraft
25-1300 MHz
*

* *Coverages are
manufacturers' claims*

These types of antennas are characterized by low cost,
1 to 3 dB gain, and "peaky" performance within the coverage.

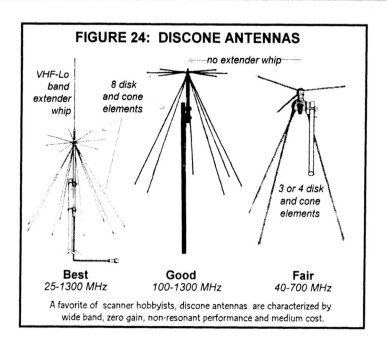

FIGURE 24: DISCONE ANTENNAS

VHF-Lo band extender whip

8 disk and cone elements

no extender whip

3 or 4 disk and cone elements

Best
25-1300 MHz

Good
100-1300 MHz

Fair
40-700 MHz

A favorite of scanner hobbyists, discone antennas are characterized by wide band, zero gain, non-resonant performance and medium cost.

Discone Antennas

Another antenna you might consider is a **discone**. The name becomes self evident when you see one. The top section of the antenna is a series of horizontal wires that resembles a disc. The lower section of the antenna consists of a series of angled wires that resembles a cone, hence **discone**. These antennas were originally designed for monitoring aircraft. However, their extremely broad bandwidth has made them a popular all-around antenna for scannists. Most discones designed for scanner monitoring can provide reasonable continuous frequency coverage from 50 MHz all the way through 1200 MHz. Some models have an additional vertical element that extends above the disc and cone portions of the antenna. Such designs allow for additional lower frequency coverage down to between 25 or 30 MHz. Such range covers the common frequencies found on most moderate-and high-priced scanning receivers. The tradeoff that you face when owning a discone is that this antenna design exhibits no significant gain and most design's performance tends to favor those frequencies above 100 MHz. Still in all, it makes a great general coverage antenna for most users.

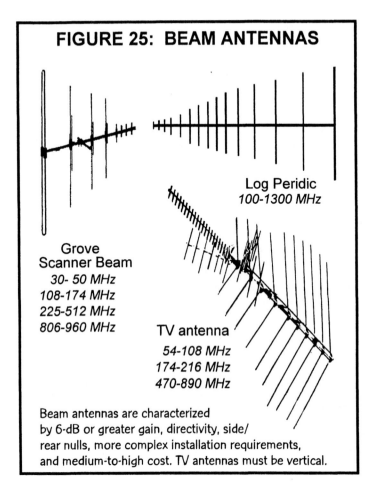

FIGURE 25: BEAM ANTENNAS

Log Peridic
100-1300 MHz

Grove
Scanner Beam
30- 50 MHz
108-174 MHz
225-512 MHz
806-960 MHz

TV antenna
54-108 MHz
174-216 MHz
470-890 MHz

Beam antennas are characterized
by 6-dB or greater gain, directivity, side/
rear nulls, more complex installation requirements,
and medium-to-high cost. TV antennas must be vertical.

Beam Antennas

The third antenna for consideration is the choice for folks who are intent on seeking out signals from as far away as conditions allow. This design is called the beam antenna. It consists of a number of vertical elements mounted along a horizontal beam. For most scanner purposes these elements will be arranged so that the antenna appears more or less like a triangle or trapezoid when observed from the side. You will see longer elements toward the back of the antenna moving progressively down to much shorter elements toward the front. The more technical name for this design is the **log periodic** beam antenna. The "point" of the triangle is the direction that signals will be heard most strongly. Many scanner monitor beam antennas can exhibit as much as 9 dB of gain in the direction in which they are pointed. Practically, this means that signals in that direction will appear to sound three times louder than from other directions. The tradeoff with this antenna design is also its biggest asset, that being *directionality*. The vertical and discone designs are **omnidirectional**, picking up signals from every direction with relatively equal aplomb. The beam antenna favors one direction over all others by several orders of magnitude. To be most effective, a beam antenna needs to be mounted with an antenna rotor to allow you to turn it in various directions. This represents an added expense but it is well worth the cost in terms of improved performance.

If you have an old TV beam antenna around the house (common if you've gone to cable) you can mount it vertically and get reasonable results on the scanner frequencies. You can fine tune the antenna further by trimming the elements according to the formulas we have discussed throughout the antenna chapters of this book.

Since antennas for VHF/UHF are relatively small and light, many monitors will eventually have several mounted for different purposes. For example, you might do your normal listening with a vertical or discone but switch over to a beam to dig out some signal that is coming in too weak to capture with the omnidirectional antenna. As you begin to experiment, you may even try out antenna designs that favor very high performance on one small portion of the spectrum that really appeals to your monitoring instincts.

Cables

Once you have figured out which antenna design you want to work with, you will need to give some serious thought to the feedline you will use to bring the signal from the antenna to your scanning receiver. All VHF/UHF antenna applications call for **coaxial cable**. Remember, this is the cable designed with a center conductor fully surrounded with an outer braided shield. There are two things you want to look for in cable design specifications. One is the percentage of **shielding** provided by that outer surrounding shield. Obviously, cable that provides 100 percent shielding is going to perform better than cable that only provides 50 percent shielding. Another performance characteristic you need to pay special attention to is the cable's **losses**. Cable specifications will usually list design losses per foot and by frequency. Remember, the higher the frequency the greater the effective loss in most cases. You will also experience losses from such actions as splicing cables together. Poorly installed cable connectors and, over time, deterioration in the cable itself caused by exposure to the elements will result in signal losses. All these losses are additive, so you have to work at your design to keep them as low as humanly possible to give your scanner every opportunity to hear everything it can.

One popular and fairly inexpensive coaxial cable you can use for receiving purposes is known as RG-6. This is common cable television (CATV) cable. It can be purchased in bulk for as little as 15 cents per foot and it exhibits very low loss across most of the frequencies of interest to the scannist. Its relatively dense braided shield is further backed by a foil conductor giving nearly 100 percent effective shielding. Its losses below runs of around 100 feet are very low compared to many other cables. Some scanning antennas come ready to accept standard CATV cable connectors. If your particular antenna does not, favoring RF style connectors, a quick trip to your local electronics supply house or a glance through a radio hobby catalog should turn up an adapter to get the job done. More expensive, but very high quality and low loss cables, are RG-11/U and Belden 9913. You can usually find these cables at radio supply houses or from catalogs that cater to amateur radio operators. These cables can be great for runs in excess of 100 feet. Avoid using "CB" style coaxial cables such as RG-58. These cables are very lossy in the VHF/UHF range and they do not usually provide adequate shielding.

Safety

Since VHF/UHF monitoring benefits from high antennas, keep in mind the most basic safety rules. Never climb higher than your ladder is designed to go. Never, never, never place an antenna or any of its support structure or feedlines where they could possibly come in contact with electrical power lines. Also, always follow the antenna manufacturer's and local electrical code suggestions for lightning and surge protection. This is made more critical by the fact that your VHF/UHF antenna system may be the highest point around and thus susceptible to lightning problems.

Typical public safety base antenna system
Note that all antennas are vertically polarized

41

VHF/UHF monitoring accessories

Not too far in the past, about the only "accessory" that you could come up with for further scanning enjoyment was an external antenna. Fortunately we live in a more enlightened age. Many interesting pieces of technology exist today to take your VHF/UHF monitoring activities to new heights (and I don't just mean your antenna height, either).

Preamplifiers

Let's start with where the signal comes in to your scanning system. If you are trying to hear weak and distant signals, you need to improve the overall **gain** of the signal as it reaches the back panel of your scanner. We have already talked about minimizing losses in the feedline and choosing the best antenna for your application (remembering that beams tend to have the highest gain in most situations). What else can you do? You can add a device known as a **preamplifier** to your antenna/feedline system. A "preamp" is a powered electrical circuit that boosts the signal to your receiver. These can be very useful in digging out weak signals so long as you are also aware of their special limitations. First, you must choose a preamplifier that is designed to support the range of frequencies that you a seeking to hear. This ability to amplify incoming signals presents us with a two-edged sword. A preamplifier will punch up those weak signals, but it will also boost nearby strong signals; this can lead to problems such as overloading your receiver and making it hard to hear *anything* well. So you will want to make sure you have the capacity to switch the preamp in and out of the antenna system. Also, preamps don't just amplify transmitted signals. They amplify all background electrical noise along with the signals. If you install a preamp at the end of your feedline where it goes into the back of your receiver, it will also amplify all of the man-made noise picked up along the coax. Not very desirable for good listening practice. For this reason, you will want a preamp that is designed to be mounted right at the antenna. This means that it will need to be water and weather-resistant. Finally, preamps are powered devices. Regardless of where they are mounted you will need to get power to them. Fortunately, modern preamps are fairly low-power devices, often working for months on a single battery, so this is not a big problem.

Overloading

We just mentioned overloading. In the world of VHF/UHF monitoring, you will probably discover you have more problems with too many signals that are too strong as opposed to too weak. A common problem in many urban areas comes from high-powered "Paging" transmitters. These are the devices that generate the signals to those little pocket pagers that everybody seems to be carrying these days. There are two accessories to consider to help with these problems. One device is known as an **attenuator**. This device essentially reduces the overall signal level coming in to your receiver to help eliminate the effects of overloading. Ah, I can hear the wheels turning out there. If a preamp boosts all signals, including unwanted ones, doesn't an attenuator reduce all signals, even the wanted ones? Yep, that's how it goes. An attenuator always represents this tradeoff. The key is to find an attenuator that allows you to adjust the amount of signal reduction to assure you wipe out the offending signal while still being able to log what you're after. By the way, many moderate-and higher-priced scanning receivers have built-in attenuation circuits. Check for this accessory when deciding on your rig.

Another less brute force way of resolving the problem with an offending strong signal is a **bandpass filter**. This is a tuned circuit that reduces incoming signals on certain frequencies while allowing desirable signals to pass. These devices are akin to the **preselectors** discussed in the shortwave section of the book. You will also find preselectors designed to work in the VHF/UHF ranges. Bandpass filters can be tunable across a small range of frequencies or they can be custom designed for a specific frequency or portion of a band.

Switching

While we are still on the subject of the antenna and feedline, remember we talked about the advantage of having more than one antenna for VHF/UHF monitoring? If you have the resources and inclination to have a multiple antenna setup, you will want to take control of this system by installing an **antenna switch**. With one of these devices you can switch between antennas and receivers rapidly. Some people have the opposite problem. They have more than one scanner but only one external antenna. You can get signals down to two or more radios from one antenna by using a standard cable television (TV) splitter. These are available at most electronics supply stores. These inexpensive devices are usually rated to cover from 70 through 500 MHz. This won't cover all of the frequencies your scanner is probably capable of, but if your primary listening falls between those two figures you might want to give it a shot. You can always remove the splitter and listen to just one receiver when you venture outside of its design limits.

Tape recording

Since most events occur over time in the VHF/UHF world, a tape recorder is essential. Monitoring enjoyment can be further enhanced with a tape recorder equipped for extended play. Another important feature is **voice activated operation** (VOX). Both of these features can be found on recorders manufactured by Viking International, 150 Executive Park Blvd. #4600, San Francisco, CA 94134

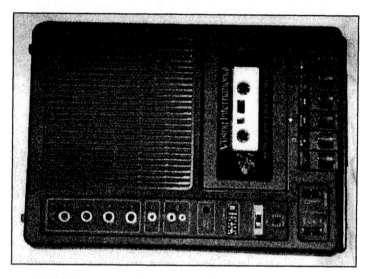

Viking International recorders are designed for the scannist

Signal display

A device that modern technology has only recently made inexpensive enough for hobby consideration is the **spectrum display unit** (SDU). This is a specialized version of a piece of electronic test equipment known as a *spectrum analyzer*. Basically an SDU will appear as a box of electronics and a visual display, usually a common computer monitor. What such devices allow the user to do is examine a band of frequencies visually. On the screen you will observe signals as peaks on a graphic display. Usually, the SDU is connected to your receiver to allow control of the tuning circuitry. This allows you to adjust the receiver to hear any signal you see on the display. The SDU also allows you to examine signals in terms of their level of strength. This can be helpful when attempting to determine the direction a signal is coming from or it can aid in the choice of antennas for monitoring the signal. If you enjoy hunting for signals, the SDU is a great tool to add to your listening post. One company currently producing an SDU at hobbyist prices is Grove Enterprises, PO Box 98, Brasstown, NC 28902-0098, (800) 432-8155.

Frequency counters

One accessory that many VHF/UHF monitors would not want to be caught without is a **frequency counter**. These devices were originally designed to be used as a piece of test equipment for radio and digital equipment servicing. But radio monitoring hobbyists are a fairly creative bunch of folks. They discovered that frequency counters could be used to discover previously unknown signals. Most frequency counters are small hand-held devices that have an antenna input and a digital display. In the presence of a strong signal, these devices can "read" the incoming signal's frequency and then display it on the unit's readout. Once you have the frequency, you simply enter it into your scanner and you're ready to listen in.

Here's a typical scenario. A new mall has opened in your area and you can't find the mall security walkie-talkie frequencies listed in any frequency resource. Just head for the mall with your trusty frequency counter and have a seat. No doubt, in a few minutes, the routine radio use within the mall will bring the needed frequency up on your counter's display. What could be easier? Well, several frequency counter manufacturers have developed counters specifically designed for the hobbyist's use. These scanners, in addition to performing their normal readout function, have the ability to store frequency information in a memory circuit. With a unit such as this, you could keep the counter in a coat pocket and just take a walk through the mall. When you get back to your car, you can extract the frequency you need from memory. Some counters are even designed to work with computer-based scanner accessories to allow you to directly enter the stored frequency information into your scanning receiver. We will talk more about these capabilities when we move on to Chapter 49.

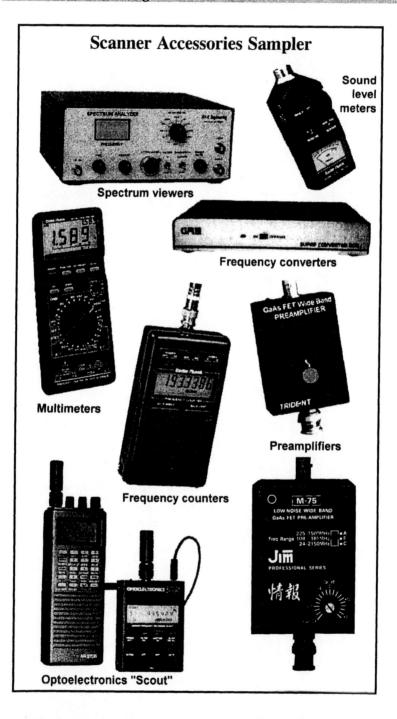

Scanner Accessories Sampler

Sound level meters

Spectrum viewers

Frequency converters

Multimeters

Preamplifiers

Frequency counters

Optoelectronics "Scout"

42

VHF/UHF propagation

In the beginning, I mentioned there were differences in propagation above 30 MHz. Scanning breaks down largely to issues of short-distance and long-distance VHF/UHF monitoring.

Short-distance monitoring

A fundamental principle of VHF/UHF communications is **line of sight**. Essentially, this means that the transmit and receive antennas usually require a clear path between the two. Most signals you will initially monitor will be local, point-to-point communications. Police and fire departments are likely to mount their main repeater antennas at points high enough for clear communications with their personnel at any point in the area. Pay attention to high buildings and water towers. You will probably see quite a few antennas lurking on the tops. The notion of *line of sight* also applies to you on the receiving end of these signals. There is no greater performance factor for your station than height of the antenna. Height yields a clear "line of sight" path to more transmitter sites.

Now, if propagation were left on this note, you could be tempted to think that the horizon is the limit of your ability to snag signals, but Mother Nature helps out. The atmosphere, having some density, causes VHF/UHF signals to bend (**refract**) beyond the optical line of sight horizon by as much as 15%-33% (*like light refracts or bends when it passes from water to air or vice versa*). If the visual line-of-sight of a transmitter antenna is 10 miles, a scanner 13 miles away might still be able to catch this signal under normal conditions. This won't always be the case, due to variations in terrain and barriers of hills, buildings, and dense vegetation, but it's a good rule of thumb.

Line of sight, on a practical level, means that your tabletop or handheld scanner, with its whip or rubber ducky antenna, should bring in most signals within a radius of 5-25 miles. If you have a good outside antenna up about 30 feet or so, you can snag signals from 50-miles or more. If your site has significant height above surrounding terrain, such as a hilltop, you will be the envy of all your scanning compadres. In almost all cases, nothing substitutes for height above average terrain for normal VHF/UHF monitoring.

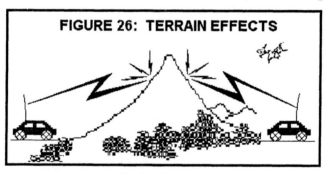

FIGURE 26: TERRAIN EFFECTS

Terrain and man-made features make the *line of sight* concept very confusing. In most cases, you won't be able to actually "see" the distant transmitting antenna, but, still have optical line of sight. That's because most vegetation and light construction (homes) are transparent to VHF-UHF radio waves. In other cases, a ridge or city skyline can completely block VHF/UHF signals, even over a short distance. To resolve such problems, VHF/UHF users employ an artificial propagation improvement called a **repeater**. Say there is a town that more or less surrounds the base of a mountain (or an artificial mountain of tall buildings). As long as two mobile transmitters stay on the same side of the mountain, they can talk with ease. But what happens when the mountain is between them?

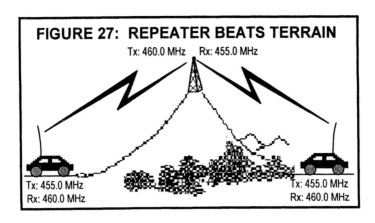

There is no more "line of sight" so communication will be essentially impossible under normal conditions. *Enter the repeater, a highly-specialized, small radio station.* A repeater **receives** signals from transmitters on one frequency and **retransmits** them on another frequency, usually with significantly increased coverage. For example, your friendly neighborhood police officer's hand-held radio may only have an output of a few watts. Operating at ground level, the officer would be hard pressed to get a signal to go more than a few blocks, much less across town. But chances are the radio will hit the police repeater system. His signals are easily received by the well-placed antennas of the repeater system and then are rebroadcast with sufficient power to cover the designated area. Repeater communication is a standard technique for most commercial and pubic safety communications.

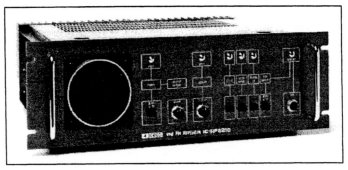

Typical repeater for the 2-m amateur band
(Photo courtesy of Icom America, Inc.)

Communications Satellites

The other form of artificial propagation you will discover is communications satellites. While this might sound overly exotic, satellite communications are becoming increasingly common. Several satellites can be heard with any moderately-priced scanner.

Some monitors even go so far as to specialize in listening in to these birds. Amateur radio operators even take advantage of communicating through satellites put in space for their exclusive use. Also, some members of the astronaut and cosmonaut corps are licensed radio amateurs who make a point of making contacts with hams worldwide from the Space Shuttle and the MIR space station.

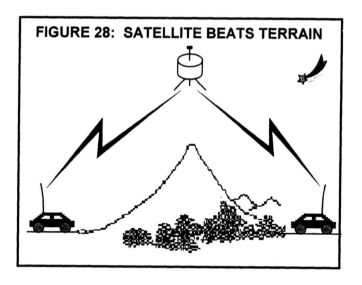

FIGURE 28: SATELLITE BEATS TERRAIN

Long-distance monitoring

While line of sight and repeater communications dominate routine listening, your goals include pushing past what is normal, toward the rare and unusual. VHF/UHF scanning offers wonderful surprises when certain propagation phenomena come into play. Making use of these additional propagation characteristics is what sets the dedicated scannist apart from those who have a scanner just to hear their local police and fire calls.

Temperature Inversions

For those who paid no attention in science class, the **troposphere** is that layer of atmosphere just below the **ionosphere** (discussed in the shortwave section). A lot of what we call "weather" occurs in the troposphere. Some of this weather helps scannists hear signals over extreme distances: **temperature inversions** occur when masses of cool air impact masses of warm air. When the masses meet, the cooler air pushes up the warm air. Effectively, a "duct" is made that traps and conducts VHF/UHF signals over distances much greater than line of sight. Signals are ducted like water through a pipe. It's possible to catch ducted signals from several states distant. Ducting can occur almost anywhere, but the phenomenon is prevalent during the summer along coastal regions and inland near stationary weather fronts. Ducting can happen in early morning after a cool night meets a hot day. Keep an ear to weather reports; you might hear some "impossible" signals. You might be listening in on your favorite local frequency and hear the regular signal interrupted by a more distant signal. If this occurs, it is a good bet you can make other rare catches. Work fast, because ducts rarely last more than a half hour.

E-layer skip

We talked about how signals in the shortwave frequency region refract off of the D, E, and F layers to bring in long distance signals. Normally, signals above 30 MHz pass through the ionosphere into space, but you know sometimes things just aren't "normal." Anyone who has played with the VHF/UHF radio spectrum for any length of time appreciates the phenomenon known as **E-skip**. At various times in the year, notably summer and the dead of winter, some areas of the E layer of the ionosphere become highly charged with ions. During such periods even signals above 30 MHz refract back to earth as if they were shortwave signals. When this occurs, it is possible to catch signals from as far as 1500 miles. These E-skip events tend to be random and brief and they rarely affect signals above approximately 150 MHz. E-skip theories are controversial and contested in the scientific community, but as the white lab coats argue over the subject, dedicated radio monitors just fill their log books with exotic entries. Just as with tropospheric ducting, your first indication of E-skip might be interference on a local channel. Again, things happen fast, so scan and log for all your might.

Auroral conditions

We talked about monitoring WWV and WWVH for propagation information in the shortwave section. Scannists with shortwave receivers should listen to the K index figures. When this figure gets up to around 4 or 5, this is an indication of increased **auroral** conditions. This can signal band openings up to about 50 MHz.

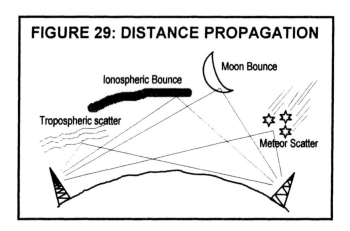

FIGURE 29: DISTANCE PROPAGATION

Meteor showers

If you want something even more exotic, check out a current astronomical almanac at a library. Meteor trails are highly ionized and enable VHF signals to travel extremely long distances. Amateur radio operators use "meteor scatter" to make very long distance contacts, a tricky but rewarding operation. Meteor signals last only a few seconds but you may be surprised at the stuff that turns up.

Moon reflections

Speaking of distance, hams use the moon as a passive reflector for VHF/UHF radio signals. Using this technique and specialized equipment, enormous terrestrial distances (up to 12,000-miles) can be covered.

As you can see, there are many ways to receive signals far in excess of "normal" expectations. As I always say, "Why be normal?" Being a radio monitor, and hearing distant stations, is much more fun.

43

VHF/UHF modes

Frequency modulation

Most VHF/UHF signals are frequency modulated (FM). Where AM involves changes of wave amplitude, FM waves stay the same size but vary in frequency. FM is most widely known for its static-free signal. Almost all scanner signals are **narrow band frequency modulation** (NFM). Scanners are designed to receive at least NFM because of its dominance in VHF/UHF applications. **Wide band frequency modulation** (WFM) is used in commercial radio and television broadcasting. If a scanner tunes the standard FM broadcast band, but doesn't have the WFM mode, the sound will be distorted. WFM mode is not particularly useful to most scannists because so few signals require it.

Amplitude modulation

If you plan to scan aircraft frequencies, you need a scanning receiver that operates in the amplitude modulation (AM) mode. Aircraft use AM instead of FM for safety reasons. You see, when two FM signals are present, the strongest one dominates and obliterates the weaker one. This is known as **capture effect**. Two AM signals can interfere with each other, but the receiver hears both. This is very important in aircraft emergencies.

This 430 MHz ham transceiver operates on FM, SSB, and CW (Photo courtesy of Icom America, Inc.)

SSB, CW, and digital

The amateur radio portions of the VHF/UHF spectrum have all of the above modes as well as **Single Sideband** (SSB), and **Morse Code** (CW). Unless you monitor exotic ham transmissions, you will not need a scanner that covers these modes. Most VHF/UHF hams operate in the FM mode. There is increasing use of **digital signal modes** by amateur, government, and commercial stations. Chapter 49 discusses the equipment necessary to listen in on these activities.

Frequency spacing

Frequency spacing is both a mode and a frequency issue in the world of scanners. Much of the VHF/UHF spectrum is regulated by frequency band plans. These are the conventions that establish "channels" and useage, for instance the public safety or the maritime portions of the spectrum. Different sections have established different channel spacings. Modern scanners are designed to automatically select the proper spacing for each established segment of the VHF/UHF bands.

44

VHF/UHF
monitoring
techniques

As I said earlier, most scannists are happy to hear a few dozen
public safety frequencies and let it go at that. But you and I are *radio
monitors*. Our nature, curiosity, and tenacity will not let us be so
content. We just *know* there's a signal out there we're missing and
we have to find it. Chasing down the hundreds or thousands of
VHF/UHF signals is the mother of all challenges. This task is much
more difficult compared to bandscanning in the mediumwave and
shortwave broadcast areas due to the nature of the signals. A
mediumwave or shortwave signal is usually constant over time.
Programming lasts for minutes or even hours. The VHF/UHF
spectrum is more like the point-to-point communication in the utility
portions of the shortwave bands. Signals are brief, sporadic, and
unpredictable. A channel can be active for only a few seconds an
hour. This makes catching signals an exercise in patience and
organization, two skills any dedicated radio monitor must master.

Memory

Your scanner's memory feature can seem complicated. You might wonder, "How do I keep track of 200 or 400 frequencies?" Granted, there are folks who can probably memorize such a list (or at least claim to) but I'm not one. For us mere mortals in the memory department, there are a few simple techniques that help keep things making sense after we've loaded up our scanner's memories.

One standby trick of the trade is good old fashioned 3" x 5" file cards. Memory channels in most units are organized in groups or "banks" of frequencies. This allows easy activation of logical groups of frequencies in the scanning sequence. It's good practice to make up a file card for each bank. Number the cards according to the bank and channel references. The first card for a scanner with 20 channels per bank, would read Bank #1 and the channels would be numbered *1,2,3 . . . 20*. The second card would read Bank #2 and the channels would be labeled *21,22,23 . . . 40*. With this system, you can record all the frequencies you routinely have preprogrammed into your scanner. Then when a signal pops up on channel 57 (*in Bank #3),* to continue our example, go to the appropriate card to help you remember what police, fire, or other station's frequency you have entered in that position. It may sound a bit elementary, but this quick and easy system helps to keep you well informed.

It's very easy to build on this simple memory aid system. First, you may discover it helpful to organize channel banks as to purpose or other logical guide. For example, you might enter police frequencies in Bank #1, fire in Bank #2, emergency medical in Bank #3, etc. Another strategy is to group your banks by towns or cities. You will develop strategies that are logical to your needs for speeding the search for ID's during emergencies or other events.

Bandscanning

Another activity you should try is bandscanning, VHF/UHF style. Assuming you have a scanner with the capability to **search** a programmed region of the VHF/UHF bands, set your scanner up to search through all of its available frequency ranges about 500 MHz at a time. You might want to vary the scan range depending on band activity; there is no hard and fast rule. This will give you the opportunity to look in places beyond any known frequencies you may already have from lists or other resources. Also, bandscanning gets you off the beaten path for signals beyond the more common police and fire calls that most people associate with scanning.

You won't often be able to immediately identify signals. This can also be complicated by multiple users per frequency. Don't get too excited about full identification yet. Just note the frequencies in your log and include some information about what you heard. After you have completed a bandscan, you can go back and enter any interesting frequencies into your scanner's memory for further analysis. You can gather enough information over time to figure out who's who and what they're up to. Figuring out what you are hearing on previously unidentified frequencies is a bit like doing detective work, and it is a lot of fun.

Don't forget that communications can be brief. Let the scanner run through the search frequencies several times to pick up all the possible signals. Also, you will want to repeat your bandscans on a regular basis. I'd shoot for once every three months or so. You will be surprised at how much things can change in such a short time. For example, a whole service might move up from VHF to UHF. Since there can be a lag in publication of new frequencies in hobby journals and magazines, your own signal hunting and analysis will keep you on top. In essence, you will develop the same discipline as mediumwave and shortwave monitors. Bandscanning makes you *know* the VHF/UHF spectrum. You can quickly become the local scanning expert if you regularly study the spectrum.

Here is a simple guideline. *NO FREQUENCY EVER GOES UNUSED FOR LONG!* Keep up with your bandscanning and you will find some amazing things. For example, I know one case where several area police departments moved en masse from VHF to UHF. *"What did they do with all their old radio equipment?"* About a month after the big change to UHF, a scan of the old VHF frequencies revealed activity. These same area police departments started using the old equipment and frequencies for inter-department drug interdiction communications. So keep in mind that changes in frequency can often turn up even more new activity if you take the time to systematically study the changes and their results.

Event-oriented monitoring

Another strategy that can bring about interesting listening is known as event-oriented monitoring. This is especially popular among those who follow sporting events, but can be applied to almost any area of VHF/UHF monitoring. Using this system, the scannist takes some time to research the likely frequencies or frequency ranges that would be active during the activity that they are interested in monitoring.

Motorsports monitoring

One area of listening that serves as a good example of how this system works is motorsports. Most professional auto racing efforts make use of ongoing radio communication between the driver and the pit crew throughout the race. Lists of the current frequencies used by the various racing teams are usually available at the race track or from other sport- or radio-hobby resources. The scannist/race fan has only to enter these frequencies into their handheld scanning receiver and don a pair of good earphones (race tracks can be loud places) to add a whole new dimension to enjoying the sport. By monitoring the track-side communications, you can gain insight into how teams compete with each other. You will also hear some fairly candid statements when things don't go as planned.

> **NASCAR racing teams have been known to speak directly to their fans in the stands, knowing that many people are monitoring their frequencies while watching the race.**

Air show monitoring

You may be interested in attending air shows at your nearby airport or military installation. Bringing along a scanner capable of tuning through the aircraft frequencies can bring a lot of fun listening your way. Performance teams such as the Blue Angels, Thunderbirds, and Golden Knights all make use of radio communication to put on their respective shows. Likewise, they need to be in communication with the tower and the other activities to make sure everything goes as it should.

Other events

But you do not have to stray far from home to apply the concepts of event-oriented scanning to your monitoring activities. Fires, car accidents, or any local event where large numbers of people gather, such as parades and concerts, also depend on radio communications in the VHF/UHF realm. Paying attention to the patterns of frequencies used at such events by the local authorities is worth keeping track of in your notes and logs. This knowledge will speed up your choice of frequencies whenever such events occur in the future. For example, knowing that your location's police and emergency medical service make use of a little-used frequency to communicate during local public events means that you might give that frequency higher priority in your listening during your neighborhood's Fourth of July parade. Local event-oriented monitoring on your part will require some scanning detective work, but it is well worth the effort.

Local geography

Another approach to getting to know what's around on your scanner is to examine your local geography. This can give you all kinds of information to help you in your scanning efforts. Remember how we said earlier that under most conditions a scanner with a rubber ducky or other simple indoor antenna was likely to give you a range of 25 miles? We also said a good outside antenna could double that distance and more. All you need to do is take this information to a map and you're ready to improve your monitoring.

You start this process by finding one or more good quality maps of your area. You are going to want more detail than you are likely to find in a road map or travel atlas. Check your local library. You are going to be drawing on these maps, so use copies if you don't want to disturb the original (no sense in getting the librarian mad at you). Figure out where on the map your monitoring post is located. Using a compass or other circle-drawing device, draw a series of concentric circles with your location at the center. What you are doing is generating a map that will indicate what towns, cities, businesses, and other scannable locations are situated within that 25-mile radius of home. By looking at a map with these "scanning circles" on it, you may discover that a few other interesting sites fall within the 25-mile radius. A closer look might reveal that one or more airports are nearby, or perhaps a military installation. Your map might also reveal a river that carries commercial shipping. Further, you might identify an industrial park or two. Again, a little radio-monitor oriented thinking will see all of these locations as scanning opportunities. Once you have become the monitoring master of all you survey within your 25-mile radius, your next challenge will be to draw your circles out to 50 miles and get a notion of what signals may be emanating from this further region. Geographic examination on a good topographical map will also reveal to you the hills, mountains, and areas of tall buildings that can impede scanning in some directions. But like they say, when the world gives you lemons, make lemonade! This map is also showing you the highest points within your region. Chances are you can get to one or more of these high points with your scanner and pick up some more distant signals while improving you monitoring in your immediate area of interest. VHF/UHF-oriented amateur radio operators often go "hilltopping" in search of ways to get their signals to travel farther. Any dedicated scanner monitor can use the same techniques for receiving signals.

Mobile monitoring

Another way to enjoy scanning is to take the show on the road. Mobile monitoring gets new signals wherever you travel. There are two cautions when mobile scanning. First, know the law before using a scanner in your car. Most states have no objections to mobile monitoring, but a few states and municipalities have restrictions. To avoid problems, check with the authorities first. The second concern is simple safety. You can't very well program a scanner and drive a car at the same time. It's a sure way to set yourself up to hear police and EMS calls directed to you as they come to extract you from your telephone-pole-kissing car. Always pull over to a safe place to make changes. I want you around to read my next book, okay?

There are a couple of ways to skin the mobile-monitoring cat. Some scannists set their receivers to scan the likely public-safety portions of the band as they cruise the highways. This gets lots of signals from various places without spending a lot of time fiddling with the scanner along the way. While this method may seem a bit haphazard, it gets results if you're not too picky on signals.

Another variation on this theme is to research the frequencies that come into range along the route of your trip. Then it is just a matter of entering the frequencies into memory and waiting for them to become active along the route. You don't need to save this technique for those long summer journeys. For example, I have a 45-minute commute to my real-world job. Along the way I pass through the coverage area of more than ten townships in three counties. The scanner I take on my daily trip is loaded with the various public safety and other interesting frequencies that usually come up along the way. At least once a month, I am alerted to a traffic problem or accident that I can then detour. Knowledge is power, and commuter scannists can have fun and get home with less traffic hassles.

Car antennas

If you intend to go mobile with your scanner monitoring, you will want to invest in a scanner antenna for your car. The same rules apply to the choice of mobile antennas as we discussed in the Antenna chapter of this section of this book. You will want to look for an antenna that covers the frequency ranges that appeal to your monitoring practices. You will want to be sure that the antenna is fed with low-loss coaxial cable. Getting cable as close to 100 percent shielding as you can find is important to reduce noise from the car's ignition. Also, if you're running your rig off the car's power, ignition noise can travel to the receiver through the power cable. The surest way to reduce this problem is to bring the power to the receiver from the car's battery by the shortest route that is safely possible. Make sure that you have a properly-installed fuse to protect your receiver. Do not attempt to utilize your car's power without reading both your car's and receiver's operating manuals to find out how to do it right. If you have any doubts, consult somebody who has professional knowledge of automobile electronics. I think your long-suffering spouse might frown on your burning out the family car's wiring harness.

A few well-placed antennas enhance mobile scanning

Remembering frequencies

One more thought before we move on. I am one of those folks who has a memory like a sieve. If you held a gun to my head and asked me the frequency of my local police department, I'd have to surrender and plead mercy due to ignorance. I just don't hold numbers well in my head (as any of my math teachers throughout my life can attest to). Does this mean I can't be an effective scannist? Not on your life! I just write everything down. This is even a good practice to get into if you are blessed with a photographic memory. Fatigue and excitement can make even the most organized monitor a bit confused. Take notes during your listening sessions and you won't lose track of things that could be important.

All of these ideas and techniques are worth a try for the beginner. As you grow in your understanding of VHF/UHF monitoring, you are sure to come up with ideas of your own that will enhance the fun of being a scannist.

45

VHF/UHF time practices

Keeping track of time usage in most places in the VHF/UHF world should be no more complicated than looking at your own watch. After all, everything is so nearby, right? Well I used the word "should" for a good reason. While in most cases this might be true, in a few instances you will have to think beyond your wristwatch to keep track of what is going on when you are monitoring. Some police, fire, and EMS organizations will make use of a 24-hour clock. This won't be the UTC time we talked about in the Shortwave section of the book. Rather you will find it will be your local time applied to a 24-hour system. All of the AM times stay the same but 1:00 PM would become 1300, 2:00 PM will be 1400, and so on up through midnight which becomes 0000. The reason such organizations keep time in this way is that they operate throughout the whole 24-hour period. This fact should once again get the wheels turning. How can this knowledge serve your monitoring activities? You will want to get to know the times that these pubic safety organization's change their shifts. Listening in at a shift change can give you lots of important data. Important daily announcements are usually made at that time. For example, you might discover that your local police force has been alerted to keep an eye out for some individual, car, or activities at a certain location. Likewise, fire departments might give information concerning drills and assemblies at shift change. Shift-change information may serve to direct your listening for the rest of the evening, especially if you want your receiver tuned to where the action is.

If you are listening in to a public safety communication and you hear the operators giving out odd-sounding information when asked for the current date, don't be alarmed and think you've slipped into some alternative universe. Some organizations use something known as a **Julian calendar** for record-keeping purposes. Even folks who use this system professionally have a lot of trouble keeping it sorted out so you can do yourself the favor of noting it in your log and then forgetting about it. This is supposed to be fun, not work!

46

Scanning clubs

Just as with the other areas of radio monitoring, clubs are a common part of the VHF/UHF world. Scanning clubs serve to help you grow in your understanding of the technology and techniques that apply to the world above 30 MHz. Clubs also serve as conduits for up-to-date frequency information exchange. It is unlikely that your local police department will publicly announce a move of its detective operations to a new channel. But this information will probably show up in your club publication soon after it is discovered. You might even be the one to share this tidbit of information with your fellow members based upon your monitoring activities. Check *Monitoring Times* for a more complete list of clubs, but here are two examples:

The All Ohio Scanner Club

The All Ohio Scanner Club (AOSC), 20 Philip Drive, New Carlisle Ohio 45344-9108, is well known among dedicated scannists. Don't let the name fool you. The AOSC is national in its scope, providing useful information for all VHF/UHF scannists. Its frequency information favors hobbyists who reside northeast of the

Mississippi River, but several columns also cover information on federal, regional, and nationally-oriented frequency information. The club's bulletin, *American Scannergram*, is loaded with frequency information, product reviews, and other specialized scanning topics. The club holds an annual meeting each summer. $3 sent to their address will bring you a sample issue and membership information.

The All Ohio Scanner Club has published the "American Scannergram" to an international audience for 15 years

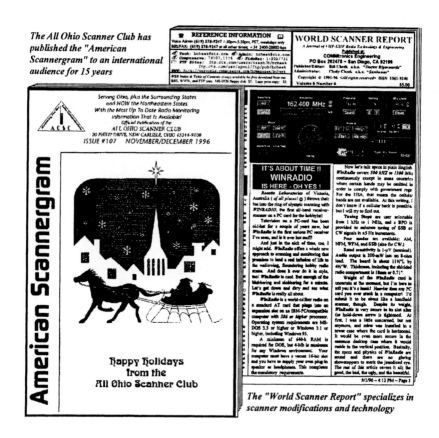

The "World Scanner Report" specializes in scanner modifications and technology

The Bearcat Radio Club

The Bearcat Radio Club produces a bimonthly publication called *National Scanning Report* that provides frequency information throughout the United States. Their publication is professionally produced and includes many feature articles on scanning techniques.

More information and a sample can be had by sending $3 to Larry Miller, Box 360, Wagontown, PA 19376; (800) 423-1331.

Local clubs — other scanning information

You are curious by nature. You have probably wondered, "How can a national scanner club meet my needs for a 50-mile radius of my site?" They can't, as a rule of thumb, but then they don't try. National scanner clubs either publish a little something for as many major metro areas as they can, or they focus on hardware and techniques and leave it to the local organizations to disseminate frequency data. Obviously, national organizations can be helpful in a general sense, but be on the lookout for local support.

There are countless numbers of regional, state, and local scanning clubs around the world. Keeping up with this ever-changing list is beyond the scope of this book, but for starters, check with your neighborhood Radio Shack dealers, ham radio and CB shops, and even electronic supply outlets for clues about local scanner clubs.

Scanning is uniquely different from other forms of monitoring in the sense that 99% of it is strictly local in nature. There never has been a proliferation of national and worldwide scanner clubs like for interests below 30 MHz. The most successful national scanner club of them all, the Radio Communications Monitoring Association (RCMA) folded and closed its presses in mid-1996 after a 20+ year tenure of service on the national scanner scene. Not that this is a sign of the times, but it underscores a focus on local resources. If your site is within a 50-mile radius of a major city or population area, you can probably find a club that wants you.

Speaking of which, just as the mediumwavers and shortwavers have their "tip networks," local scanner groups use *"notification networks,"* organized along the same lines. The more avant-garde notification networks employ pagers to facilitate tipping their members "en masse" when all hell breaks loose on the scanner. Again, check your local radio stores and electronics outlets for information about scanner clubs and these notification networks.

Some of the magazines discussed in the next chapter publish news of the more prominent regional clubs and activities/events.

The following regional clubs and organizations are listed for your convenience, but this is only a representative sampling and probably doesn't scratch the surface of what's out there.

CARMA (Chicago Area Radio Monitoring Association)
PO Box 2681, Glenview IL 60025
BBS: (708) 852-1292
SASE for sample newsletter

New England Scanner Group
PO Box 1024
Derry, NH 03038

Incident Command Page of Arizona
`We Page With Frequency'
Web http://www.disaster.net/icp
Web http://www.getnet.com/~n7qyk

Mountain NewsNet
James Richardson
PO Box 4488
Estes Park, CO 80517-4488
(970) 586-4325 (business)
(970) 586-4357 (fax)
jimfun@aol.com
Web http://www.frii.com/~rmedic/

North Carolina Tarheel Scanner/Shortwave Listeners Group
Contact: Marshall Sherard
Address: 4612 Winterlochen Dr.
Raleigh, NC 27603-3868
(919) 505-2145 or (919)-773-1695
Fax: (919)-773-1695
E-Mail: KE4ZNR@aol.com
        MSherard@aol.com
Net every Monday Night 9:30PM on 146.640 MHz., Meet on 3rd Mon ea month. Call For Loc & Time.

Fire Notification Network of Michigan
Garry Watts
PO Box 1312
Warren, MI 48090-1312
(810) 772-4423 Voice
(810) 773-1346 Fax
firenet@usa.net
Web http://users.aol.com/
              fnnm/www/homepage.html
Michigan alphanumeric paging network for notifications of breaking news events. Customizable for fire, police, EMS, severe weather alerts, national fire, general news and sports.

Miami Valley DX Club
Dave Hammer
BOX 292132
Columbus OH 43229
Voice Phone:614-848-9410
E-Mail:MVDXC@juno.com
Web http://www.anarc.org/mvdxc/
All Bands with emphasis on SW and Scanning

FirePage (r)
Tom Kravitz
POB 1307
Culver City, CA 90232
(310) 838-1436
(310) 838-4495
E-Mail: Mpage@netcom.com
Web http://ourworld.compuserve.com/
              homepages/Media_Page
Fire-notification network covering the state of California.

47

VHF/UHF
publications

Commercial publications that are useful to the VHF/UHF
monitor abound. This aspect of the radio-monitoring hobby is
perhaps the best supported by magazines and newsletters. First we
will review two magazines we have already talked about in detail in
other sections of this book, and then we will learn about two more
publications dedicated exclusively to the world of scanning.

Monitoring Times

Monitoring Times (*MT*), a monthly, has its roots in the scanning
world. Its publisher, Bob Grove, has served as the mentor for many
of those who claim any expertise in the radio hobby, including your
humble author. While *MT* remains a great general-purpose radio-
monitoring magazine, it is also a standout in the area of VHF/UHF
matters. The editorial staff includes several well-respected experts in
the scanning hobby. The magazine supports two primary VHF/UHF
oriented columns, "Scanning Report" and "Scanning Equipment."
These columns are only the beginning. Many other regular columns

and feature articles are of specific interest to hobbyists who monitor the world above 30 MHz. If, in the course of your VHF/UHF monitoring experience, you become interested in listening in on satellite communications, *MT* has a sister publication called *Satellite Times*, devoted exclusively to this aspect of the radio monitoring hobby. *Monitoring Times* and *Satellite Times* can be ordered by calling 1-800-438-8155 or by writing to P.O. Box 98, 300 S. Hwy 64 West, Brasstown, NC 28902. E-mail: **mt@grove.net** Their Web site is at: **http://www/grove.net**

Popular Communications

Popular Communications (often referred to as *PopComm*) is another monthly magazine with a general radio interest but a deep commitment to the scanning aspect of the hobby. The main scanning column, "Scanning VHF/UHF," is further supported by many feature articles and columns that discuss many practical aspects of the radio art. The magazine also has columns devoted to satellite and emergency services communications for anyone interested in these areas. Of the several commercial radio-monitoring magazines, you are most likely to find *PopComm* on regular newsstands. Otherwise, you can call 1-800-853-9797 or write to 76 North Broadway, Hicksville, NY 11801 for more information.

U.S. Scanner News

U.S. Scanner News (USSN) is a monthly magazine devoted exclusively the VHF/UHF scanning hobby. Each month its experienced editorial staff cranks out a series of columns devoted to the most interesting aspects of the monitoring world above 30 MHz. Among these columns is "Frequency Forum" that publishes those frequency changes and updates that have occurred in the previous 30 days. This is one of the easiest ways to get information that can be used to update any existing frequency resources you may be using. For more information about *USSN*, call 1-503-230-6999 or write to P.O. Box 14923, Portland, Oregon 97214.

World Scanner Report

The *World Scanner Report* is a monthly newsletter published by Bill Cheek, who is widely known for his expertise in radio modifications and for his ability to show non-technical hobbyists how to improve and wring out maximum performance and features from their scanner equipment. Bill is the author of several books on scanner modifications, including his latest, *The Ultimate Scanner,* published by Index Publishing Group.

Cheek launched the *World Scanner Report* in January, 1991, to keep scannists updated between releases of his scanner hacking books. The *WSR* charts new turf on the scanner technical scene and helps with logical decisions about radio hardware purchases. Price/info about the *WSR* for a SASE to COMMtronics Engineering, PO Box 262478, San Diego, CA 92196 or e-mail: **bcheek@cts.com** Web: **http://ourworld.compuserve.com/homepages/bcheek**

We will talk more about scanner modifications in Chapter 49.

National Scanning Report

National Scanning Report, edited by Larry Miller, is published bimonthly and is the official publication of the Bearcat Radio Club, but is devoted to all aspects of scanning. Write to Box 360, Wagontown, PA 19376.

> One of the important aspects of all magazines mentioned above is that they are great resources for suppliers of radio-hobby equipment, publications, and software. In any issue you will find dozens of dealers, publishers, and computer services that are able to serve your advancement in the arts and sciences of radio.

Budgets

This is a good spot to discuss budgetary issues again. It would be easy enough to run out and subscribe to all these magazines and to join a couple of clubs. Your annual expenses for publications could easily reach the neighborhood of half the cost of a new scanning receiver. My suggestion is that you sample an issue or two of each of the magazines, newsletters, and club bulletins that suit your fancy, and then buy in after you are sure they will hold your interest.

Maybe after a year or so your interests or needs will have changed. Then it's time to take another look at what is out there to support your monitoring efforts. Having said this, I still encourage you to read all you can about VHF/UHF monitoring. Knowledge is power, and up-to-date information is worth its weight in scanners.

You can have the most expensive scanning receiver ever designed, but if you haven't read the club journal that listed all those frequency changes in your area, your receiver will tune static while someone with a moderately-priced rig and a head full of information fills their log with new signals. Keep these budgetary ideas in mind as we move on to frequency resources in the next chapter.

48

VHF/UHF frequency resources

Frequency acquisition is the primary area of research for the scannist. Knowing which frequencies carry what information is usually the key to success in the world above 30 MHz. For example, you hear a siren in the distance. You could just grab your scanner and randomly search the public safety frequencies. This *might* bring up a useful "hit" before the event is over. Things happen fast in the VHF/UHF world. A better technique would be to have previously entered into the scanner's memory banks the specific local police, fire, and EMS frequencies. This should bring the signal and all the excitement of monitoring it to you quickly and with no trouble. Prior planning and frequency research is the key.

You may wonder how to get frequency information for your area. This is decidedly a simple task. Several resources will help you fill your memories with desired frequencies to get the job done.

Police Call

One of the first books that most beginning scanners acquire is the nationally-known Gene Hughes' *Police Call Radio Guide*, published by Hollins Radio Data. Don't let the title fool you. It contains a great deal more information besides police frequencies. You will find all aspects of public safety, public service, and government listed in its pages as well. This frequency book is published in separate editions for each of nine regions throughout the United States. It is updated and revised yearly and can be found at most Radio Shack stores as well as from other radio-hobby suppliers. The book is organized to use in two ways. First you can simply look up the locations that are within 25 or so miles from where you live and start plugging in any of the listed frequencies you find interesting. This should get your scanner really active in short order. But you have also professed to being a radio monitor. You will not be satisfied with just these signals. *Police Call* is cross-referenced by frequency. This means you can also go on the hunt for new signals and then try to determine what you're hearing by checking the information listed for those "unknown" frequencies. If you travel to another part of the country beyond that covered by your regional guide, just pop into a Radio Shack store in that area and get the edition to support your new listening environment. You can obtain more information about this resource by writing Hollins Radio Data, P.O Box 35002, Los Angeles, California, 90035; be sure to enclose a SASE.

Scanner Master

Scanner Master Corporation publishes a series of regional frequency guides for many areas of the United States. Their listings go beyond public safety to include frequencies for shopping malls, amusement parks, and other areas of regional interest. If you are a frequent traveler, get your hands on the latest edition of their big book called *Monitor America*. This book gives you thousands of frequencies for all types of listening throughout the country. If you take your scanner on vacations or on business trips, it's an important tool. You can find out about the complete Scanner Master publication line at (800) 722-6701 or write to Scanner Master Corporation, PO Box 428, Newton Highlands, MA 02161.

Transportation resources

As you develop in your monitoring practices, you may find that specialized areas of listening appeal to you that go beyond the more general frequency lists. Other sources are available to help you out, especially in the area of transportation monitoring.

For people who are hooked on monitoring the aircraft frequencies, Robert Corbin's *Aeronautical Frequency Directory* will give you everything you need to hear all the activity in, out, and around your local airports. In addition to information on commercial air traffic, this book also provides a great deal of useful information about military air operations. You can obtain more information by calling (603) 432-2615 or by writing Official Scanner Guide, P.O. Box 525, Londonderry, NH 03053.

If your interest is trains instead of planes, you may want a copy of *The Compendium of American Railroad Frequencies* by Gary L. Sturm and Mark J. Ladgraf. This book is exhaustive in its coverage of rail related frequencies and information about the use of the various frequencies you might hear. Contact Kalmback Publishing Co., 21027 Crossroads Circle, PO Box 1612, Waukesha, WI 53187.

"Secret" frequencies

In addition to traditional information, *SCANNERS & Secret Frequencies* by Henry L. Eisenson takes its readers off the beaten path to examine some of the more unusual things you can find with your scanner. This is the first book I've seen that has a comprehensive list of "televangelist" frequencies. This highly regarded book is available from Index Publishing Group, Inc., 3368 Governor Drive, Suite 273, San Diego, CA 92122, (800-546-6707).

Federal frequencies

Most of these frequency guides are a bit light on federal government frequencies. If you are a "fed hunter," consider a copy of the *Master Frequency File* by James E. Tunnel and Robert Kelty. This book covers every aspect of the federal government for FBI, DEA, and Secret Service through Customs, and the National Park

Service among many others. More information on this book can be obtained from Tab Books, Division of McGraw-Hill, Inc., Blue Ridge Summit, PA 17294-0850.

CD ROM

With the growth of the personal computer market, the latest trend for frequency resources makes use of CD ROM technology with listings of the Federal Communications Commission's license database on a single CD ROM that can be managed by common database-management software. Such resources are exhaustive but the data tends to be in a fairly raw form. Still, once you've got a handle on their use, CD ROM frequency lists supplemented by updates from club and commercial publications are the ultimate tool for the scannist. Two major resources publishing CD ROMs at this time are Grove Enterprises, 1-800-438-8155, P.O Box 98, Brasstown, NC 28902 and PerCon Corporation, 716-386-6015, 4906 Maple Springs / Ellery Road, Bemus Point, NY 14712

Other frequency resources

Don't neglect your monitoring skills. There is no substitute for tuning around and finding things yourself. These resources are just guides to get you started. Join a scanner club for an important source of frequency data. Never underestimate the direct approach. I've had reasonable success just asking people using radios for their frequency. There's no law against asking, and even if your request is sometimes refused, you will still find quite a few who will share their information. Keep records of your discoveries and share with others. This is how our collective frequency database really grows.

> At a radio convention in a major hotel, one persistent scannist followed a security guard for over an hour trying to monitor the guard's handheld frequency. When I saw this, I walked up to the guard and asked for the frequency. He obliged with a smile. Asking seldom hurts.

49

Computers and VHF/UHF monitoring

Probably no other aspect of the radio monitoring hobby has benefited more from the development of the personal computer than has scanning. If you are not already someone who is computer oriented, VHF/UHF radio monitoring might just be your chosen pathway to the world of computers and the Internet.

Frequency management

Compared to most other aspects of the radio-monitoring hobby, people interested in the world above 30 MHz tend to require the ability to manage large numbers of frequencies that are often subject to change. The process of frequency management alone can serve to demonstrate the practical uses of a personal computer for the scannist. Modern database management software programs can

261

allow you to keep track of *whose* signals are on *what* frequencies. Such programs also allow you to massage the entered data to group frequencies in whatever patterns you may choose. For example, working from an existing database you have either entered yourself or purchased, let's say you want to have a list of all the fire department frequencies within your county. Usually a few taps of the keys can separate this information out, and give you just the list you need. Many scannists use database-management software to maintain their logs. For this purpose you can even purchase several programs specifically designed for the radio hobbyist. However, if you are reasonably computer oriented or have a desire to learn, any of the general-purpose database management programs such as dBase, FoxPro, Alpha, or Access can be found on many business and personal computers, and can be used to maintain your log. These programs can do the job even better, because they allow you to customize your entries to suit your specific needs and desires. But frequency management is only one small way that a computer can be used around the listening post.

Word processing

Just as with the other areas of radio monitoring, the personal computer can be used with word-processing software to enhance your hobby activities. Information lists, confirmation letters, and submissions to your club bulletins or commercial magazines all become so much simpler when done on a personal computer. Further, you can all but eliminate the piles of paper from your listening post. All of my personal scanning logs, letters, and columns written on the subject for various magazines reside on a couple of 3½ inch computer disks. If I needed to keep all this stuff in hard copy, I would probably fill a drawer or two of a traditional file cabinet. Rapid access to information is the key to VHF/UHF monitoring, and the home computer speeds things up a great deal.

The Internet

Scannists have made great use of the Internet, the worldwide system of computer information. VHF/UHF radio monitors communicate with one another, sharing information and frequencies far more rapidly than even the best club newsletter or commercial magazine can. Getting "online" as a scannist is just one more way that computers can further enhance your monitoring activities.

Receiver/computer linkage

Now I'm going to let you in on the really exciting stuff about computers and scanning. First, "How do scanners work?" Think for a minute. When we talk about working with our scanner, what terms do we use? Programming? Memory? Information entry? Hmmm. Where else do you find these terms? That's right, these are "computer" terms. By now you may have figured out that modern scanning receivers use microprocessors (computers) to do their thing. Now let's take this idea a bit further: can a scanning receiver be tied to a computer? **Yes, yes, yes**, and with amazing and exciting results too! Personal computers with special software can give you almost infinite control of your scanner. You can rapidly enter frequencies from a database into your scanner's memories. You can search and store new frequencies and analyze them on the fly. You can operate your scanning receiver from a remote location, allowing you to reduce the length of signal-attenuating coaxial cables. All these features and more are possible when computers and scanners are wired up to one another.

Until recently, computer control was an exclusive of high-end scanning receivers with a data port as a factory feature. In recent years, several companies developed add-on circuits that enhance the capabilities of moderately-priced scanners. These aftermarket boards and software essentially add the data port feature the manufacturers left out for dozens of advanced features that would be impossible without the personal computer. With a computer interface, you can enjoy the same level of computer control as you might find on a scanning receiver costing many hundreds of dollars more. Currently, two companies provide retrofit hardware needed to bring this feature to your scanner, COMMtronics Engineering, PO Box 262478, San Diego, CA 92196-2478 (**bcheek@cts.com**) and Optoelectronics, 5821 NE 14th Ave., Ft. Lauderdale, FL 33334.

It can be scary to think about opening up your scanner and making modifications. But if you have an interest in computer control and are capable of or are willing to learn basic electronics construction techniques, you can create a scanner that will marry up to your home computer and save a lot of money along the way.

Packet radio

If you've tuned the VHF/UHF amateur bands, you probably heard odd noises. Some of these are **packet radio**, a digital transmission mode that allows text and program files to be sent over the air. It's possible to listen in on the fun by connecting your scanner to a personal computer through a device known as a **terminal node controller** (TNC). Usually the connections are no more complicated than plugging one cable into your scanner's speaker output jack and another cable into your computer's RS-232 port. With this equipment and a little bit of software you can hear what all the fuss is about. Better still, you are halfway to having your own packet radio station if you choose to get your amateur radio license.

An efficient and functional amateur packet station

Scannists who monitor satellites use software developed to help them track the various bird's orbits and movements. Some hobbyists even use their personal computers to aim directional antennas to keep the satellite's signals coming through their receivers.

In the Shortwave section we talked about the problems of noise interference from computers. This problem is often less obvious at the VHF/UHF frequency region. While you may not detect the noise directly on the audio coming out of your receiver, often the computer-generated interference will create "birdies" and lock-ups when your receiver is scanning. You need to take the same steps as you would in the lower-band situations. Experiment with physically moving the receiver and the scanner's proximity to one another.

You can have lots of fun with a scanning receiver. You can have even more fun with a scanning receiver and a personal computer. You can become the absolute master of every signal within sniffing distance of your antennas if you have a scanning receiver connected to a personal computer. As you advance in your understanding and practice of monitoring above 30 MHz, you just might find yourself becoming a computer person along the way.

Radio on a computer

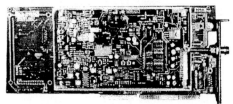

Rosetta Laboratories of Victoria, Australia entered the computer/scanner market as this book went to press with *WinRadio* and a slight twist; it comes on an AT card that plugs into an expansion slot of a PC/clone computer with 386 or better processor. *WinRadio* tunes 500 kHz-1300 MHz and is operated like a conventional radio, except from the keyboard, mouse, and monitor of your computer. *WinRadio* does it all: mediumwave, shortwave, and VHF/UHF with AM, NFM, WFM, SSB, and CW modes. *WinRadio* scans, searches, and uses files for memory in 1000-channel blocks. WWW: **http://www.winradio.net.au**; E-mail: **info@winradio.net.au** For other contact and sourcing info, see *Appendix 1.*

Computer Interfaces for Scanners

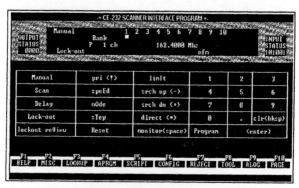

COMMtronics Engineering
CE-232 Scanner/Computer Interface

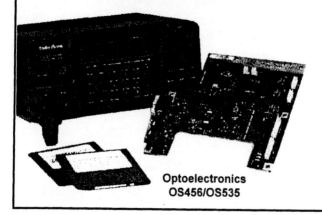

Optoelectronics
OS456/OS535

50

Advancing your scanning skills

The lament of the scannist is "*So many frequencies, so little time.*" Things happen fast and then they're gone. Being on the right frequency at the right time is a skill that scannists must hone throughout their entire scanning experience. To get a handle on what's going on depends on ability to quickly size up an event and to get the right frequencies activated in the scanner's memory. This skill comes with time, but you can help yourself get up to speed.

Here is a non-scanning example of what I mean using a scanner-related sport, auto racing. Experienced racing teams keep files of information for each track they race. Files contain information discovered over the years about what has worked at that particular track, and what hasn't: information about how to set up the car, and how to instruct the driver, all with the goal of improving performance on each track at which they race. Now stretch that analogy with me a bit to the world of radio monitoring. You probably keep a logbook of frequencies and resources. Also keep a logbook of events and how you monitor them. You may even spend some time brainstorming how to approach events that *could* come

267

up. Just as fire departments have practice drills, you will want to work out some practice scanning problems on paper. All of this can be kept as reference material for future scanning opportunities.

Say there was a fire in your local industrial park. Activity will be noted on police and fire frequencies. But where else would traffic pop up? Think through the problem while poring over a frequency list. Does that industrial park or the business have any frequencies of their own? What fire departments are likely to provide mutual aid if things get out of hand? Anticipate the frequencies needed to hear everything that is likely to happen. Now dig a little deeper and think about other problems that might pervade such an emergency. You might find action on public service, gas, or electric company frequencies. An industrial fire might require their action. A major industrial fire is likely to be a news event. What TV, radio, and print media organizations are likely to show up? What frequencies do they use to cover such events? Industrial fires sometimes involve environmentally-sensitive materials. What frequencies are the state and federal environmental-protection services likely to use in response to an industrial accident? Take the problem to its extremes and then reel in your thinking with a priority list of frequencies for whenever you encounter such problems.

In addition to brainstorming hypothetical situations, you can also take notes during your "real time" listening sessions. This is great for helping you rule out frequencies that are not likely to be useful. For instance, you might discover after monitoring several fires in your area that the fire department uses one frequency for administrative purposes that never seems to come alive during emergencies. Once you have noted this bit of data in your logs, you will know to give that frequency a lower priority when following a fire. If I monitor something exciting or out of the ordinary, I sometimes develop a few notes as a sort of after-action report to myself. When the heat of the hunt has passed, I can review these notes for how well the session went and how I might improve my performance for the next such event.

If you have scannist friends who belong to a scanner club, compare notes and scenarios with each other to see what worked and what didn't. In spite of how serious all this dress rehearsal stuff sounds, it is a lot of fun. And a few minutes of planning can make your listening that much more fun as you hit those frequencies right on the nose with the assurance of an expert.

Modifications

The curiosity that leads folks to enjoy radio monitoring often takes a side trip into the world of radio modifications and improvement. This drive toward "hardware hacking" seems to be especially strong in the VHF/UHF world. Initially, it seems this pattern of behavior was brought about by the urge to modify receivers to hear cellular phone frequencies. Now that this listening in on phones is technically illegal and newer receivers are difficult if not impossible to modify to bring in the cellular frequencies, you might think that this urge to tear into scanning receivers would have abated. Not so. The modern electronic circuitry that can be found in most scanners lends itself to various forms of improvement for either general or specific listening activities. Despite some claims to the contrary, modern equipment is actually somewhat less difficult to fiddle with than older gear. Even a beginner, with a good information resource and a few simple tools, can perform many modifications that will take their scanning receiver well beyond the expectations of its original design.

There are a few warnings that should be issued at this point before you go melting any solder:

   (1)   Modification of a receiver will no doubt void any warranties or service agreements you may have with the manufacturer or supplier. If you are uncomfortable with this notion, you should hold off on ripping the covers off until any warranties have expired

   (2)   Before you attempt any modification, check its "pedigree." Just because somebody posts a note on the Internet that removing Resistor #256 from a particular receiver's circuit will enhance performance doesn't make it true. Depend on well-documented modifications from reliably published resources. I've already made mention of Bill Cheek's scanner modification books and newsletter. Thoroughly documented resources such as these, where you are sure the modification

has been tested out by a competent professional, remain the safest way to go when getting into hardware hacking. It also doesn't hurt to check with other users who may have tried the modification themselves. You may want to cruise the Internet or computer BBS systems to see if anyone has noted any unpredicted side effects to a modification, but shy away from BBS systems as a primary modification source. People can post anything and then just disappear.

(3)   Even though most scanner modifications will occur on low-voltage circuits, you are often dealing with receivers that have power supplies energized by house current. The stuff that comes out of your wall sockets can *kill* if not treated with respect and safety. *NEVER* work on a receiver that is energized. If you have no experience with being safe around electricity, by all means leave the covers on and take the time to learn about electrical safety from a reliable resource before you attempt any modification.

Once you are comfortable with the above concerns, you can then embark on the adventure of bringing out the best your scanning receiver has to offer. Working on electronic equipment can be a bit anxiety-producing until you get the hang of things. My suggestion is you seek out a more experienced individual to help you get started. Maybe somebody in your local scanner club or a neighborhood amateur radio operator will be willing to help you learn your way around inside your receiver. Most of what I've learned about electronics over the years came from the kindness of dozens of radio hobbyists who were more than willing to share their knowledge with someone less experienced than themselves. If you take the time to look around, you will find this spirit still exists in abundance in the radio hobby.

I like to think I hold the world's record for voiding a warranty. While waiting for a Radio Shack sales person to ring up my sale on a PRO-2005 scanner, I took the cover off the rig to clip the frequency restoration diode. I also began several other "improvements" to the receiver while waiting. Hardware hacking is fun no matter where you do it.

Antenna spotting

Another activity that attracts a lot of interest among VHF/UHF monitors is "antenna spotting." If you are ever in a crowd of people, you will be able to pick out the dedicated scannists in the crowd. They will be the ones looking up at the tops of buildings and water towers, in search of elusive antenna structures. You see, a spotted antenna can often be the indication of a previously unknown frequency and signal resource. Once you have read through a few more extensive texts on radio antenna design and construction, you will become adept at figuring out the users of antennas and even the likely band of frequencies you can check out to see what signals are coming from it. From this point a quick scanner sweep or a glance at a frequency counter and you've got another bit of information to share with the folks in your scanner club. Some people even take this process a step further by making use of special surveillance receivers designed to allow you to monitor a nearby signal without needing to know the specific frequency.

One such device is sold under the name of The Interceptor by Optoelectronics, 5821 N.E. 14th Avenue, Ft. Lauderdale, FL 33334. With the addition of this type of surveillance receiver to your monitoring tool kit, along with a frequency counter and your regular scanning receiver, you are prepared to discover signals no matter where they might turn up.

Law enforcement activities

If you get seriously involved in the more surveillance-oriented aspects of radio monitoring, you may even discover such signals as police stake-outs or even hidden "wire" microphones being used in the course of a law enforcement investigation. My suggestion to you, if you do run across some quasi-covert activity, is to STAY OUT OF IT!!! Law enforcement's job is hard enough without a hobbyist blowing its cover.

Public safety participation

Something else that appeals to many scannists is a strong desire to get "on the other side of the scanner." Why just listen in on the action when you can be part of it? In most parts of the country, volunteer fire departments, emergency medical service teams, and agencies such as the Red Cross are looking for folks just like yourself who can help their community. Even if you don't have the urge to hang off the back of a fire engine, please consider contributing to your local public-safety organization either financially or with your time. They surely could use a hand and you can think of it as the rent you pay for all of those exciting signals you get to monitor.

> While staying in a motel on a business trip, I relaxed by doing a bit of scanner monitoring. Much to my surprise I discovered that the local police were conducting a stakeout of a room near mine. Needless to say, I didn't get much sleep that night.

As you continue to grow in your monitoring skills you will make many exciting discoveries about the world around you. Mix a little study into your scanning schemes and you will discover even more than you imagined.

51

Are we really still having fun???

I certainly hope so. Monitoring the world above 30 MHz puts us in closer contact with humanity. Most VHF/UHF signals are about close-by events. Some of what we hear has an impact on our lives and well-being. Heady stuff! Some people focus on the dark side of the scanner events: fire, crime, and disaster. If your listening is limited to these, you may be in the proverbial rut. *Radio monitors* experience something above and beyond most scanner owners. They get a larger picture of the world around them. Sure, we hear about fires, but we also hear about the heroism of our firefighters who save lives and property. We hear about crime, but we also see law enforcement at work around the clock preventing society's predators from taking over. We hear of disaster. We also hear how people from everywhere come together to put it right for the victims.

If listening to your scanner convinces you the world is a dark, terrible place, you've missed out on part of the story. Move out from those *"cop and robber"* frequencies to hear the world at work. Hear people like yourself going about their daily tasks. Tune the search and rescue frequencies to hear people helping people. Tune the business and transportation bands to hear people serving people. Discover that most people are okay. Some of these less-exciting signals are a sign that the world is just fine in spite of the alluring negativity. You can't help but get a positive outlook when you listen beyond the police and fire signals.

If you read the hobby radio press, you may see convincing opinion that VHF/UHF radio may soon become inaccessible due to efforts to scramble and encode signals. While this is a possibility, scrambling techniques have been around for decades and have never really taken over. Scrambling introduces a higher, costlier technology which in a way defeats the purpose of radio. You may discover that signals will change, but there will always be signals to monitor. And there will *always* be dedicated, tenacious radio monitors figuring out how to listen in on the world above 30 MHz.

From the mediumwave, shortwave, and VHF/UHF sections, you learned the basics that will enable you to monitor most radio signals that come your way. Use this information to gain access to the distant world and the one as close as next door. You are able to see the world around you through very different eyes because you have the means to listen with very different "ears." You hear the world through radio monitoring.

Have we covered some turf? Yep. Have we covered everything you want or need to know? *I hope not!* This book just gently pushes in the right direction. If you don't have more questions now than when you opened this book, I probably didn't do my job. Radio monitoring is about insatiable curiosity coupled with the patience and tenacity to persist to a goal. I hope these notions have been awakened so you can embark upon the journey of a lifetime. At the risk of sounding like a Zen mystic, think of this book as a finger pointing the way. You should have more questions now than when you started, but I think you'll discover these questions are more focused and clear. Seeking answers will advance you further along the radio path. This adventure only begins. Remember, radio monitoring is a hobby that lasts a lifetime.

But I'm not finished with you just yet. I need to let you know of a few other interesting areas for the compleat radio monitor. I also want to give you a few thoughts that apply to all of the radio monitoring areas we have discussed.

52

Amateur radio

I'll bet you figured you wouldn't get out of this book without a sales pitch for ham radio. As an Extra Class amateur with almost 20 years under my belt, you could say I'm committed to ham radio. Amateur radio is one of those subjects to which most every radio monitor gives a few thoughts at some point. The urge to *transmit* as well as to *receive* radio signals can be compelling. If you've done any radio monitoring of the shortwave or VHF/UHF bands, you've no doubt run across hams doing their thing and wondered about it. You probably heard hams talking to each other around the world. You may have heard amateurs in public service activities such as message handling or supportive communications during disasters. You've no doubt heard stories about amateur radio operators making the difference in communications when the telephone networks have broken down. Emergency communications is one of the reasons amateur radio operators have retained their privileges and bandwidth over the years. Among the vital public safety and service missions are thousands of opportunities to have fun and learn about radio electronics and communication along the way.

Okay, so just what is the bottom line here? Just what does it take to become an Amateur Radio Operator? Basically there are no significant limits to becoming a ham. There are no barriers of age, sex, or even technical ability. You'll find operators as young as five years and a few so old that they were around when there was nothing *but* Amateur Radio. You need not fear being a beginner because Amateur Radio privileges are grouped, more or less, according to ability and experience. I say "more or less," because many qualified hams drop in with the novices to help them come up to speed in this exciting facet of the radio hobby.

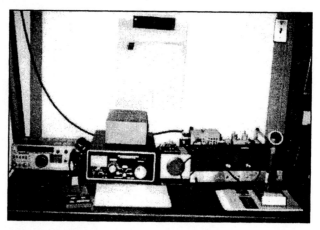

Typical amateur station ready to reach out to the world

All it takes to enjoy ham radio is an Amateur Radio License. You qualify for a ham license by taking a series of examinations known as **elements**. The class of Amateur Radio License for which you qualify depends upon which elements you pass. Don't let the idea of a whole passel of tests scare you. The element structure just organizes the process of developing as an Amateur Radio Operator into easy to master, bite-sized segments. There are five classes of Amateur Radio Licenses, each with different privileges: **Novice, Technician, General, Advanced**, and **Extra**. We will concentrate on the two essentially entry-level license classes.

Novice license

The Novice class license is the land of beginners! Novices have access to a chunk of frequencies where they can develop good operating practices that will make higher class license privileges that much more fun. Novices can operate in the HF portion of the radio spectrum on 3675-3725 kHz in the 80-meter band, 7100-7150 kHz in the 40-meter band, and 21,100-21,200 kHz in the 15-meter band using **continuous wave telegraphy** (CW). This means International Morse Code. For the Novice exam, you have to demonstrate the ability to understand the International Morse Code at five words per minute. Don't let this scare you. With a good instructional tape, most folks can get to this nominal speed with only a few weeks of study. Novices get single sideband (SSB) voice privileges in the 10-meter HF band on 28,300-28,500 kHz. Novices also have the right to use certain VHF frequencies. Novices can operate from 222.1-223.91 MHz with 25 watts and from 1270-1295 MHz with 5 watts. On VHF you can experiment with all authorized modes of amateur communication including radioteletype (RTTY) and the "digital" modes such as packet radio. These VHF privileges serve to wet your whistle for all that is available under the Technician class of license.

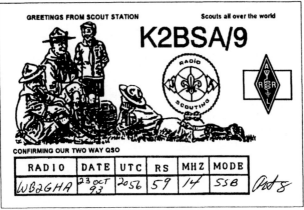

The Boy Scouts have a merit badge for amateur radio

Technician license

Just as the Novice license appeals to the shortwave monitor, the Technician ticket especially excites VHF/UHF monitors. Technician Class privileges give you access to all amateur radio frequencies and operating modes *above* 50 MHz. A Tech ticket allows 50-54 MHz, 144-148 MHz, 222-225 MHz, 420-450 MHz, 902-928 MHz, and 1240-1300 MHz. And to make the pot even sweeter, you can use all kinds of modes of communication including code, voice, data, RTTY, and TV. You can even skip the Novice exam and take the **No-Code** Technician's test. This gets you the above privileges but blocks your access to the shortwave ham frequencies until you have passed the code test. The No-Code Tech ticket is the most popular ham license going these days because most people can study and pass the exam in a very short time.

> **The "No-Code" Technician's class license has become the most issued license in the U.S.**

Study resources

Many companies have produced study guides to help you prepare for whatever level of ham license your little old heart desires. The best single-source study guide for beginners, covering both the Novice and the Code-Free Technician Class licenses, is *Now You're Talking: Discover the World of Ham Radio*, published by The American Radio Relay League (ARRL), 225 Main Street, Newington, CT 06111. This book will work you through all the theory and regulations, including the complete question pools for the multiple choice questions that will make up the test elements you will need to pass. You will also discover techniques for learning the Morse code required for the Novice exam. It even gives you solid information about how to get on the air once you pass your exam.

Back when I was first licensed, I had to trudge over to the Philadelphia Customs House where the FCC had a field office for ham exams. A few years ago, the **Volunteer Examiners Program** was established, where qualified hams give the exams for all classes of licenses. You can obtain a list of Volunteer Examining Teams in your area from The ARRL's Educational Activities Department at the above listed address. If you are a scannist, you might listen in to your area ham repeaters. Sometimes, organized nets give information about upcoming test sessions. Also, most hamfests have testing as part of the program.

Join the fun! Amateur Radio lets you add to the monitoring possibilities out there. You will congregate with an entire world of folks dedicated to hobby radio. Wade on in, the water ain't deep!

53

Other monitoring opportunities

When we covered mediumwave, shortwave, and VHF/UHF monitoring, we left out a few areas that appeal to some folks. These areas fill in the gaps in the "DC to Daylight" picture of radio monitoring.

Longwave

In our tour of the radio frequency spectrum toward the beginning of the book, we made mention of the various radio signals that could be heard in the Low Frequency (LF) and Very Low Frequency (VLF) portion of the radio spectrum. Monitoring the world below about 530 kHz often gets lost in and among all the opportunities that higher frequencies provide. This is largely because many receivers do not tune below the mediumwave band. If you have equipment that will let you listen down in the low frequencies, take some time to tune around. You will find many signals worth logging, especially beacons and "Lowfers," those hobbyists transmitting in the region between 160 and 190 kHz. This area of monitoring has enough adherents to have spawned its own club. **The Longwave Club of America** (LCWA) can be contacted by sending an SASE to 45 Wildflower Road, Levittown, PA 19057. They publish a monthly journal called *The Lowdown*.

FM and TV DXing

Just as many people enjoy listening in on mediumwave and shortwave broadcasting stations, there are those who enjoy trying to log long-distance signals in the standard FM broadcast band of 76–108 MHz. These folks take advantage of some of the unique propagation phenomena we discussed in the VHF/UHF section of the book to hear signals beyond what comes over most people's car radios. Similarly, there are people who attempt to capture the audio and video signals associated with television broadcasting. Watching a TV program from several states away without the aid of cable is quite an accomplishment. Folks interested in these activities have banded together to form the **Worldwide TV/FM DXers Association** (WTFDA). You can find out more information about how to grab these signals by writing to them at PO Box 514, Buffalo, NY 14205-0514. They will send you a sample of their journal *VHF-UHF Digest* for $2.

For FM Broadcast, FM-DX, and SCS freaks, there is the time-honored *FM ATLAS*, a regularly updated book that lists in-depth details of US, Canada, and Mexico FM broadcast stations and their programming. *FM ATLAS* has maps of each state with FM stations in each city clearly shown. The book also has two cross-reference sections where FM stations are listed by state/city and by frequency! *FMedia*, a monthly newsletter, whets appetite and interest between revisions of the book. Info:

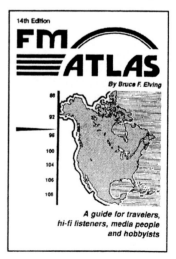

FM Atlas Publishing
Box 336
Esko, MN 55733-0336
218-879-7676   Fax: 218-879-8333

54

Setting up a monitoring post

Seeking out your initial monitoring space requires that you find the best place in the house that gives you privacy, power (electrical, that is), and access to the outside of the house. Dedicated radio people sometimes drag themselves out of bed in the wee small hours of the morning to hear something that is not there any other time of the day or night. Likewise, listening during "normal" hours can be frustrating if other members of your clan disturb you just as that ID of Radio Freedonia you have been seeking for six months comes over the air. What you want is two-way privacy.

Every house I have ever been in seems to have a corner where things that are seldom used get piled up. Stuff that is always sort of out of sight and out of mind. This might be a good place to begin your hunt for a station location. If your children have yet to discover the joys of owning copious amounts of clothing, there may be a closet that can be turned into an ideal monitoring spot. Basements, if they are not too damp and dreary, are also popular places for a listening post. A great out-of-the way place is a corner in a spare or guest room that is not in regular use. Avoid attics and garages unless they are sufficiently climate controlled. If you want to wear two pairs of long johns while enjoying a hobby take up ice fishing!

Part of thinking out your monitoring location is going to be your direct access to electrical power for your receivers and any other equipment you draw into the fray. Of course these needs will be different for each person. At the minimum, you will need to have one grounded outlet to plug in the receiver. If you have more than a few accessories you will want to consider one of the many power strips that are available on the market. These are especially useful because they are usually fused and have a master power switch. Better-quality power strips also provide protection against line voltage surges, further protecting your investment. Be careful not to exceed the recommended capacity for either the power strip or the wall outlet. *If you have any questions concerning your household power and its use consult a licensed electrician!*

You will want to have plenty of light to make reading and writing possible without eyestrain. Depending on which frequencies you frequent, try to stay away from fluorescent lighting. Fluorescents can cause unwanted interference. Stick with incandescent light bulbs for best performance. Try to locate the lighting so that it does not cast shadows when you are reading and writing.

One of the first signs that someone has actually decided if they enjoy the radio monitoring hobby or not is when they put up an outside antenna. Planning for first and future antenna installations should be part of your listening-post project. Easy access to the outside world for antenna lead-ins is not as tricky as it sounds. Usually the easiest route outside is through a window. You will also need to give consideration to how you might run your antenna ground connections. You may want to review our discussion of antenna/ground entry points in the shortwave section of this book as part of your planning.

My old Zen Master used to say "It is not the cup that performs the task, it is the space within it." The same holds true with most rooms. Now that you have zoned in on your shack location you will want to give some thought to making the space most useful.

After you have picked out a desk and chair that suit your needs you will want to plan for maximum use of the remaining space for that research and record keeping stuff we talked about earlier. The first law of great listening post design is, *You can never have enough shelves*, closely followed by my second law, *You can never have enough filing cabinets.* A couple of shelves right over your receivers will hold all of those important frequency-reference materials. A two-or four-drawer file cabinet is just the ticket for storing articles, log sheets, and other record-keeping materials.

If you want to make things as efficient as possible there is a neat "Human Engineering" experiment you can perform before you even drive a single nail. Put a chair in the spot you plan to sit during your monitoring sessions. First look straight ahead. Assuming that your receivers are arrayed on your table top (tilted upward to avoid neck strain of course), a point at eye level is the ideal place to install a shelf for your most needed reference materials. Now, from the same sitting position move your dominant hand around the desk top and room space. In the Martial Arts this is known as the **dynamic circle**. Everything within the immediate reach of your hand can be controlled quickly. This is an important notion in office design, not just in street fights. From your operating position you can now envision the most likely locations for desk, drawers, file cabinets, shelves, and switches.

Shelving comes in all shapes and sizes and can be had for very reasonable prices. Shop around a few hardware stores and lumber yards until you find what is right for your location. The only proviso I would make is that you make sure the shelving is sturdy and firmly installed. Nothing can ruin a DX session like a load of books pouring down on your head.

SCANNING TOOLS

55

Safety

It's probable that I will repeat myself on a few points in this section. *Good!* Safety is an important aspect of the hobby that bears frequent repeating. Don't forget to refer to the safety directions for antenna installation that we discussed in the shortwave section of the book. But in addition to putting up antennas, there are other safety considerations surrounding electricity that cannot be ignored.

Robert A. Heinlein once said "The only crime is ignorance and the only sentence is death." Rather strong stuff but relatively true. The radio monitoring hobby does place us in close proximity to electricity. Lack of knowledge in this area has the potential to kill. We tend to take the stuff that comes out of the wall sockets for granted. But it doesn't take a whole lot of electricity to kill someone. Over the years, building and electrical codes have changed and improved so that the electricity in our homes can't harm us so long as we treat it with respect.

The first safety consideration should be to *Never override existing safety devices*. Let's assume your receiver came with a three-prong grounded plug. The purpose of the third prong is to allow any electrical short in your equipment to route itself safely to ground. If you removed this third prong or if you used an adapter to allow you to plug your gear into an old-fashioned two pronged socket, you have set yourself up for disaster. Without a direct path to ground, any stray electricity will look for another route. That route could be through your body and that might have fatal results.

Another maxim of radio monitoring safety is *Keep your hands where they belong*. Most consumer electrical devices come with a warning clearly stating something on the order of: "TO PREVENT FIRE OR SHOCK HAZARD, DO NOT REMOVE THIS CABINET COVER. NO USER SERVICEABLE PARTS INSIDE. REFER SERVICING TO QUALIFIED PERSONNEL." The reason they put these warnings on equipment is that it is easy for an inexperienced person to get zapped by poking around inside electrical equipment. Further, equipment that isn't even turned on or plugged in can still give you a jolt. This is because electrical components known as capacitors can essentially hold a charge until something comes along and discharges them. If that something happens to be your finger, you will find yourself in trouble. Treat all electrical equipment and circuits as if they are energized at all times. If you have no training in electronics, take your equipment to someone who does before you try to fiddle with it yourself.

Another thing to keep in mind is to *Keep an eye on your power cords*. The power cord coming out of the back of most electrical devices is usually subject to all manner of injustices during its life span. Don't let your cord run under rugs or where folks can step on it or trip over it. Also, don't dangle it down behind your listening post desk in such a way that it is subject to abrasions from sharp corners of drawers or doors. A frayed cord can lead to a fire-producing situation. You don't want to hear your own address come up on your scanner now, do you?

Never forget that *If a fuse blows, there is a reason*. I have never run across a fuse or circuit breaker that died of old age. If your receiver's fuse blows or you trip your house's circuit breaker, make a thorough investigation into the possible cause before proceeding to replace the fuse or reset the breaker. Most causes are fairly obvious. I am sure we have all tripped a circuit breaker a few times by having too many electrical devices running at once. This becomes almost too common in a world now dominated by multi-outlet power strips. The solution is simple. You need to turn a few things off. You may also need to consult with an electrician to determine whether you need to change or increase your electrical service to prevent such problems from happening.

But some fuse and breaker problems are a little less obvious. If you cannot come up with a logical conclusion, seek the help of someone with a stronger background in electronics. The reason for this is that one of the causes of fuse-and circuit-breaker tripping is a short circuit. Replacing a fuse or resetting the breaker against a short can set you up for a situation where you might experience an electrical shock. Never try to resolve your problems by installing a larger fuse or breaker than indicated by the manufacture. This may mask an ongoing dangerous situation. Whenever in doubt get the advice of a trained professional technician or electrician.

As I said in the antenna chapter of the shortwave section, *Do not string antenna wires or lead-in wires near electrical services.* Pay close attention to where your house wiring, telephone lines, and cable TV system come into your house. You do not want any antennas or feedlines strung in such a way that they could come in contact with these services, especially your house wiring. Keep in mind you must also consider the possibility of an antenna coming loose and one end falling across such wiring. A little bit of caution and forethought in this area can prevent a lot of heartache later.

Don't forget the need to manage the "Big" electricity, that being *lightning*. Lightning is devastating. None of your radio equipment can survive a direct lightning hit. Furthermore, even hits some distance away from your listening post can still generate voltages that will severely damage your equipment. Obviously, it is highly desirable to keep lightning outside the home. The only protection from lightning is to completely disconnect all of your equipment from all antennas and power lines when a storm is in your area. Outside antennas should be grounded as discussed in the shortwave antenna chapter, to allow for safe routing of any static discharge. Devices sold as lightning arrestors or surge protectors are helpful against transient voltages from lightning strikes at some distance from your antennas, but nothing can withstand the full force of a direct lightning hit. The best course of action remains taking steps to keep the lightning effects totally outside of your listening post. The few minutes it takes to unhook and rehook your equipment and antennas with each use is nothing compared to the weeks of waiting for your receiver to return from the repair center. You should also check your local building and fire codes to assure any efforts you make in the area of lightning protection are within the rules.

Remember that knowledge is power. You only have to fear the above mentioned dangers if you ignore those safety practices that will protect you from harm. Always take the extra time to think things out so you will do them right and safely. When in doubt, defer to someone with more experience.

56

Logs and confirmation

Throughout human history, everywhere in the world, people who study other people have learned some basic truths about Homo Sapiens. We all like to collect things and we all like to keep score of what we collect. These primal desires are at the very root of the radio monitoring hobby. Listening to radio signals, whether down the block or around the world, becomes more meaningful when you keep track of what you have heard. **Logging** the many signals we seek will usually lead to a desire to **confirm** the loggings. This is most often done by seeking some form of verification from the originating station. Yes, folks, even in the hobby world there is paperwork! But it need not be drudgery. Once you get the hang of things, it can be as much fun as listening.

Your log is a permanent record of your listening experiences and accomplishments. But it serves more functions than simply giving you something to look back over 25 years from now. A log book is the repository of all the data you will need in order to go about the process of seeking confirmation of what you have heard. Consequently, some very specific information needs to hit the pages of your log book. More on this in a minute.

Also, a well-kept log gives strong clues to your listening patterns. Few of us are both too rich and too single. Family responsibilities and the need to eke out a living make it impossible to listen all the time. Even if you are independently wealthy, you have to sleep! A log book and a few frequency and program resources will help you get a handle on your listening needs. You have to be listening when the stations are transmitting. If your log indicates you do all of your listening on weekends and Radio Freedonia only broadcasts on Tuesdays, you'll have to reorder your life to make the catch.

Both sides of a typical QSL card ~ one side provides logging data

A log book is a matter of personal style. It does not need to be very fancy. I know of one world-class amateur radio DXer who has kept his records on common stenographer tablets for over twenty years. Commercial log books are available from many radio hobby suppliers. My personal preference, over the years, has been to design my own log sheets, making copies via any nearby photocopy machine. As computers continue to permeate the radio hobby world, it is possible to use any commercial database management program to make a nifty electronic "paperless" log. What you choose to use as a log book is secondary to the information you collect in it. Whatever you choose, use something that is sturdy enough to stand up over the years. Also, make sure you make your entries *large*. Your older eyes will thank you when you try to show your log to your great-grandchildren.

Assuming I have now convinced you of the need to keep a log, we can take a look at the kind of basic information you will want to keep track of. As you progress in the monitoring hobby, you will probably come up with additional information you will want to keep in your log.

Log entries

If you are a shortwave listener, you will want to keep track of your loggings based upon **UTC** (Coordinated Universal Time) **time** and **date**. UTC, also known as GMT (Greenwich Meridian Time), is the time standard commonly recognized around the radio world. I also log local time and date. If you are a scanner monitor, you may want to track time using a 24-hour format if your local public services use this format over the air. Don't forget to log the time you began listening and stopped listening to a station.

Make an entry space for the station's **frequency**. This might sound a bit obvious, but it can give you additional insight into a station's patterns. For example, noting a station on a frequency other than the one published in a magazine or club journal might be something you will want to share with your monitoring colleagues

The call sign or station name is important, So is the country, state, city, or transmitter location, depending on your listening habits.

Make note of the **language**. Getting this right is important to a verification. Also, make note of the gender of the announcer.

Next is where you record what is important to the station you are monitoring. Radio hobbyists have adopted a system called **SINPO** over the years. SINPO stands for a signal's <u>S</u>trength, <u>I</u>nterference, atmospheric <u>N</u>oise, <u>P</u>ropagation disturbance (fading), and <u>O</u>verall merit. Rating a signal from 1 to 5 in each category, with 5 excellent and 1 poor, this system will allow you to write a thorough verification report. Use the SINPO code in your log but *not* in your report. The person who reads your letter is not likely to be a radio hobbyist and may not have the faintest notion of what SINPO means.

Most stations want to know more than how well you enjoyed their programming. A few notes on the **program content** and how it appealed to you will go a long way in "greasing the wheels" that can lead to a verification.

Astute hobbyists use more than one **receiver or antenna** during a monitoring session. Make note of what you used to catch each station. Over the years it is fun to see just how many things you heard with each receiver or antenna you have owned.

You can use your log to **keep track of the verification process**. Make note of those stations to which you sent confirmation letters; which ones responded; and how quickly. Also, if your log shows that a station hasn't responded in a reasonable period (usually around six months), you can send another report.

Confirmations

Now let's look at how to use the information in your log to seek confirmation of what you have heard from the transmitting stations. This process is commonly called QSLing. QSL is the old Morse code operator's abbreviation for "I am acknowledging receipt." Most of what I'll spell out here applies to seeking verification from shortwave stations but can also be applied to mediumwave and VHF/UHF monitoring practices.

Sending out a verification report is a win-win situation. The station receiving the report gets an idea about its listeners and how well it is being heard. In return, the dedicated listener gets a QSL card or a letter verifying reception. QSL collections are one aspect of the radio hobby that non-hobbyists also seem to enjoy. The journey to a vast QSL collection begins with a few simple rules and a little wrestling with the international postal service.

Always **type or print your reports**. Cursive writing is confusing enough if you are from the same country as the writer. Include your name, full address with no abbreviations, Zip-code, and finish your address with "The United States of America."

Next put in the **date** you monitored the station. Always spell out the month's name completely because 2/1/97 will be interpreted as either February first or January second depending on who reads your letter in what country.

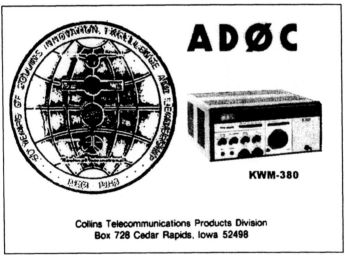

Many amateur radio operators welcome reports from listeners

Include the **name and address of the station**. This is not simply common letter-writing practice. It is also another step in letting the reader know exactly who you heard. Many broadcasters will give their mailing address over the air. If not, addresses can be found in hobby books such as *The World Radio TV Handbook* and *Passport to Worldband Radio* that we discussed earlier in the shortwave section of the book. You will find some mediumwave addresses in the *WRTH* but will probably need to depend on club publications for the latest information. Most VHF/UHF addresses are as near as your local telephone book. You just have to do some hunting.

With the top of the letter completed, you can move on to the **"meat and potatoes" information**. Start with a short paragraph informing the station that you are a radio hobbyist who enjoys listening to stations from all over the world, especially those at some distance away. You can even mention a little bit about yourself (e.g., I am an 18-year-old engineering student).

In the second paragraph, restate the date and give the time you began to monitor the station. On international verifications, always use UTC as this is commonly respected. If you want to be sure to get the point across, use a world time chart (one can be found in any good world atlas) to include the local time at the broadcaster's location. With mediumwave and VHF/UHF signals, be sure to refer to the time in terms that the transmitting station is using. Don't forget to account for time zone differences where appropriate.

In the next few paragraphs, report exactly what you heard with as much detail as possible. Now you can see why you kept all of that information in your log. Pay attention to program content. Was the announcer male or female? Was the broadcast just in English or did you hear another language used? Make note of the time that programs begin and end, making special mention of station identifications, sign-ons, and sign-offs. The exception to this rule is when you are attempting to verify any government or military station. In such cases only refer to the transmitted content in general terms to avoid difficulties. You might use generalizations such as "Routine police traffic" or "Communication between aircraft call sign 'Top Gun' and 'Ft. Swampy' tower." This way you're getting the point across without creating problems with police and military agencies concerning confidentiality.

When writing to broadcast stations, tell the station what you thought of their programming. Tell them what you enjoyed. Tell them if you liked the music. If you learned something new about the station or the country, let them know. Nothing can sour the QSL process faster than making the station operators feel that the only reason you listened was to get a QSL card. Also, if you disagree with a station's politics or religious perspective, keep it out of your letter. Arguing politics or religion will work against a favorable reply.

After you have reported what you heard and why you liked it, include a solid paragraph about the conditions. This is where you take into account the signal information you logged using the SINPO code. Stations especially appreciate information about any interfering signals. You may want to include a few lines describing the receiving equipment and antennas you used as well as your local weather conditions.

Finally, after you have given the station operators information that may be of use to them, include a closing paragraph *politely* asking for verification of your report. Don't get pushy! No station is under any obligation to write back. State what you have enclosed in terms of return postage and thank the station profusely for their time and kind consideration.

Sending a letter around the world is different from sending one across town. There are pitfalls all along the pathway of the international postal system. A few tips will help to **make sure your letter arrives at its destination**.

Use common (unfancy) air mail stamps and use plain envelopes that are thick enough that the contents cannot be examined when the letter is held up to a light bulb. This will prevent your letter from being sidetracked by some unscrupulous postal clerk.

On the outside of the envelope, include the complete addresses, yours and the station's. Do not use abbreviations. Spell out the full name of the countries. Avoid using improper country names. Mail sent to The Peoples Republic of China addressed as *Red China* just gets tossed in the dust bin. Don't be offensive. The people on the other end are just as proud of their homeland as you are of yours.

You can include return postage either by using International Reply Coupons (IRCs), available from most large post offices, or by enclosing mint (unused) stamps from the country you are writing to. Mint stamps can be purchased through any store catering to stamp collectors (check your yellow pages) or you can use The DXers Stamp Service, operated by William J. Plum, 12 Glen Road, Flemington, NJ 08822. An SASE will bring you a list of Bill's offerings.

Don't forget to include return postage with your reports sent within the United States too. If you are sending a verification to a pirate broadcasting station, it is likely that they are moving their mail through a "mail drop." These stations will usually announce their address and practices within their programming so pay close attention. The most common practice is a request for three first class units of postage. This covers the cost of moving your letter and the QSL card through the mail drop.

When sending verification letters to non-broadcast stations, you can never be sure of their full understanding of the confirmation process. It is out of the ordinary for most shortwave utility or VHF/UHF public safety stations. What hobbyists do in such cases is to include with their letter a "Prepared Card." This is simply a common postcard on which you have prewritten the information you want verified such as date, frequency, and call sign of the station you logged. In the body of your letter, politely request that the station review the information and if they find it to be correct, sign the card and return it to you. Don't forget to include appropriate postage for the card's return. In many cases, the prepared card is the only way to achieve a confirmation, so this is an important strategy.

> **A great school project that always attracts attention is a world map surrounded by QSL cards with string leading from each card to its respective country.**

Entire books have been written on the subject of QSLing. These basic hints will get you started down the road to the first 50 or so countries. Remember to be patient because mail moving around the world can take some time. The longest period I waited for one rare QSL card was *three years*! However, after about six months, you may want to consider sending another report. If possible, log the station again and send a newer report.

57

Listening, the law, and common sense

One question I hear from beginners all the time is: "Is it legal to listen in on radio communications?" The best answer is yes, with a few specific exceptions. Most of what you will hear on your receiver is covered under the 1934 Communications Act. Under this Federal law, radio-monitoring hobbyists are considered third-party listeners. The law states that *third-party listeners are prohibited from revealing the content of intercepted transmissions and cannot use the information they gain from listening for personal profit*. So basically, you can listen in to most things, but revealing the content is a no-no under the law. As you read through club and hobby publications, you will get a sense for what is considered acceptable when reporting loggings. For example, when logging a military aircraft, you might report that you heard "Flight XYZ" at a certain time but it would be inappropriate to go into detail about the pilot's discussion with the tower or other planes.

Electronic Communications Privacy Act (ECPA)

There are a small number of signals to which listening constitutes a violation of existing law. These signals are protected communications under a body of law known to most people as the ECPA. In 1986 the Federal Government produced the laws behind the **Electronic Communications Privacy Act (ECPA)**. Without getting into all the legal mumbo-jumbo, the law "protects" cellular telephone owners and users from being monitored by making it against the law to listen to the frequencies that contain cellular transmissions. Further, in 1994 it became illegal to manufacture or import receivers capable of hearing these frequencies. Later provisions under the law restricted listening in on cordless phone signals, and it has been long against the rules to monitor the **Subsidiary Carrier Service (SCS)** signals without the express permission of the transmitting station. SCS signals are programming transmitted on a portion of an FM common transmission. These signals normally require a special receiver or modification to an existing receiver to be detected. So let me state most clearly . . . **DO NOT LISTEN TO CELLULAR TELEPHONE FREQUENCIES! DO NOT LISTEN TO CORDLESS TELEPHONE FREQUENCIES! DO NOT LISTEN TO SCA SIGNALS WITHOUT PERMISSION! NO, NO, NO, A THOUSAND TIMES NO!!! YOU ARE A BAD LITTLE RADIO MONITOR IF YOU COMMIT THESE ILLEGAL ACTS!!! IF YOU COME NEAR A RECEIVER AND ONE OF THESE FREQUENCIES CAN BE HEARD, QUICKLY COVER YOUR EARS AND BEGIN TO SHOUT LOUDLY AS YOU RUN FROM THE ROOM. I AM GIVING YOU FAIR WARNING THAT THESE ARE UNSPEAKABLE CRIMES.**

More than a few folks think that the ECPA is the silliest law put on the books since Prohibition and the 55 mile-an-hour speed limit. But, still in all, it remains the law and somewhere in the world there are folks who take this law very seriously, mostly people with a lot of money invested in the cellular radio service. Practically, if your receiver just happened to accidentally stumble across a cellular signal, there is no ECPA Radio Police to smash your door down and drag you off to the pokey. As you move on in the hobby you will even discover that this law has a few holes in it that you can drive a large truck through. For example, it is legal to own a receiver made before 1994 that covers these frequencies; you're just not allowed to listen to those frequencies that are prohibited by the ECPA. Also, ironically the standard UHF television channels 80 through 83 tune right through the cellular frequency bands. Also, it is perfectly legal to construct signal adapters to make post-ECPA receivers cover these banned frequencies, just as it is not against the law to build a circuit to decode SCS signals. *You just aren't supposed to listen to anything you hear.*

If your Momma raised you right, you will just avoid these signals and continue to enjoy all the other exciting things there are to hear in the radio frequency spectrum. However, if your Momma didn't raise a fool, you'll see how silly and essentially unenforceable certain aspects of this law are.

Still, it is important that you know the law is on the books. Protect yourself as you, your conscience, and any legal counsel see fit. I'm sorry I had to burden you with this so early on in your hobby experience but this world has become a crazy place.

Scanner use or possession

Mostly applicable to the VHF and UHF regions are some state and local laws that restrict *where you can be* when you do your listening. Several state and local governments place restrictions on carrying receivers capable of monitoring police communications in a car. Still other states reserve the right to restrict ownership of such receivers by people previously convicted of crimes.

Somehow I think, if I was planning to commit a major felony, I'd be more worried about whether or not the police caught me with illegal weapons than with an illegal radio. Still, the laws exist and need to be respected.

At the time of this writing Florida, Indiana, Kentucky, Michigan, Minnesota, New York, Rhode Island, and South Dakota currently have some restrictions on scanner use or possession. This body of law is constantly changing.

It is my strong suggestion that, if you have any doubts, contact your local authorities to make sure your actions do not take you over the line. Remember, "Ignorance of the law is no excuse."

Scams

Since we have just demonstrated that there is very little common sense in some aspects of radio law, let's take a look at some common-sense aspects of the radio hobby. As you grow in the hobby, you will, hopefully, be taking advantage of the opportunity to correspond with folks all around the world as well as with folks who share your interest in radio monitoring. There is a famous tale that goes around radio monitoring circles. One dedicated shortwave listener sent a verification report to a South American station, but instead of getting the QSL card he expected, he received a phone call from a nearby airport. It seems he was listed as the sponsor for immigration of an individual from the country to which he had sent the report. The immigrant was at the airport and waiting for him. While this is an exceedingly rare case, you might encounter attempts to "exchange currency" and other questionable practices. Likewise, not everybody you encounter in the radio hobby is likely to be somebody you want to show up your doorstep unannounced. Do yourself the favor of protecting your privacy by making use of a post office box for your correspondence. A PO Box is cheap insulation against the more eccentric things the world can bring your way.

Pirate stations

Speaking of post office boxes, I mentioned earlier that most "pirate" broadcasters use post office boxes established as "mail drops." Pirate operators are willingly and knowingly broadcasting without legal authority to do so. So they use mail drops as a way of protecting their identity from the authorities. There are no laws against listening to pirates, and there are no laws against writing to them for confirmation by way of mail drops.

> **Currently active pirate mail drops include (among others):**
> - Box 452, Wellsville, NY 14895
> - Box 109, Blue Ridge Summit, PA 17214
> - Box 28413, Providence, RI 02908
> - Box 293, Merlin, Ont N0P 1W0 Canada

Attractive QSL cards are common in the pirate radio world

Summary

I don't think there is anything in this chapter that should keep you awake nights worrying that the radio cops will descend on your house and drag you away in chains. Basically, if you don't tell anybody what you hear or make use of what you hear for financial gain or some other nefarious purpose, you can happily monitor without fear of retribution. The laws that restrict listening cover a very small segment of all the things you can hear in the radio-frequency spectrum. Knowing and following the law will not put a serious damper on all the fun that radio monitoring has to offer.

MORE CLASSICS OF AN ERA

Sony ICF 2010   LF-MF-HF

Sony ICF-PRO80
LF-MF-HF-VHF

Sony ICF-SW1   LF-MF-HF

Realistic DX-440   LF-MF-HF

Uniden BC-760XLT   VHF-UHF

Uniden BC-220XLT
VHF-UHF

These classics of 1988-1994 may be good buys now

Epilogue

I'm still having fun!

Now that you're ready to put this book down and get on with the greatest hobby in the world, I thought it might be fun to give you a look at some of the listening I've done during the process of pulling the book together. Keep in mind that a lot of the time I would have spent at the dials was given over to creating this book.

In the mediumwave world, I've logged a number of new stations including WJDM, 1660 kHz, Elizabeth, NJ. This is one of the first stations to set up on the newly expanded AM broadcast band. Also, the Army Broadcasting Service conducted a test of a transportable AM broadcast station for use in support of our troops in Bosnia. Using various call signs, including KTRK, this station appeared for a brief time on 1670 kHz. By the time you read this, other stations will be both testing and operating between 1610 and 1700 kHz. Because this new upper portion of the band is still uncluttered, monitors can hear stations without much interference.

The shortwave bands were as active as ever. Both broadcast and utility frequencies brought about many interesting loggings surrounding the events in Bosnia. The movement of American troops to stabilize the region and the various involved countries' reactions have made for great listening.

I also added several new "spy number" stations to my logs. Many people thought these would more or less disappear from the airwaves with the end of the Cold War. But these stations continue to transmit their intriguing signals, keeping hobbyists speculating as to their purpose. The hurricane season this summer included several storms that had an affect on shortwave broadcasting. The Caribbean relay stations of several major broadcasters were damaged by storms, resulting in shifts in programming to other relay sites. This made for some very different loggings.

Being a confirmed tinker, I've enjoyed monitoring the shortwave bands lately using several relatively "low-tech" receivers based on designs from the early days of radio. It's amazing how much monitoring can be done using old-fashioned regenerative receiver designs.

Pirate radio activity continues to be high. A record number of stations have been on the air this year and this allowed me to add quite a few to my log book. The several "shut downs" of the Federal government during the 1995/1996 budget debates brought about a lot of activity based on the assumption that nobody was being paid to catch them.

On the amateur radio HF bands, I added a number of new countries to my overall totals. This was done in the course of doing what most hams enjoy best. Tuning around and talking with new friends from all over the world. I'm beginning to devote more of my ham time to operating in the CW mode. With commercial and government operations shifting away from the International Morse Code, CW operating becomes more of an art. I want to do my bit to make sure it doesn't become a lost art.

In the VHF/UHF region, several nearby public safety communications systems moved their operations to new frequencies. Locating and logging the new channels and their use was made more efficient by using a scanning receiver interfaced with a personal computer by way of a COMMtronics Engineering CE-232 interface. A task that once might have taken several weeks took only a few hours using computer assistance.

More traditional methods were applied to track down the operating frequencies of my local shopping mall's new security service. Monitoring a frequency counter's readout while having an enjoyable lunch at one of the mall's restaurants gave me all the information I needed to keep my frequency lists up to date.

The northeast corridor of the United States was knocked back by a record snow storm. The VHF/UHF bands were hopping with activity as my home state and several surrounding states declared emergencies. In addition to great monitoring, I had the opportunity to get in on the action by participating in local **Amateur Radio Emergency Service** (ARES) and **Radio Amateur Civil Emergency Service** (RACES) nets, assisting in communication and transportation with area emergency service agencies.

Get the idea? There is always something new and different to monitor, no matter what portion of the radio frequency spectrum you choose to tune. Its time for you to begin. Have fun! I'm looking forward to reading *your* book in a few years.

A frequency-finding frequency counter

A CD-ROM can hold about 650-Mb of data!

Appendix 1

Equipment suppliers

This Appendix contains a list of companies that deal in radio monitoring equipment. It is by no means exhaustive, but it should serve to get you started. If you subscribe to one or more of the commercial hobby publications mentioned in this book, you will discover many more resources. I have not personally dealt with all the companies in this list, so you will need to evaluate each and determine how they might best serve your needs.

ABC Communications
17550 15th Avenue NE
Seattle, WA 98108
(800) 558-0411
Receivers, scanners, and accessories

ABC Electronics
1210 S. Garey Avenue
Pomona, CA 91766
Receivers, scanners, and accessories

Ace Communications
10707 East 106th St.
Fishers, IN 46038
(800) 445-7717
Receivers, scanners, and accessories

ACK Radio Supply
3101 4th Avenue South
Birmingham, AL 35233
Receivers, scanners, and accessories

Alpha Delta Communications, Inc.
1232 East Broadway, Suite 210
Tempe, AZ 85282
(602) 966-2200
SW antennas, lightning protection

Amateur Electronic Supply
5710 W. Good Hope Rd.
Milwaukee, WI 53225
(800) 558-0411
Amateur radio transceivers; receivers

Japan Radio Co. Ltd.
430 Park Ave. (2nd Floor)
New York, NY 10022
(212) 355-1180
Shortwave receivers

Associated Radio Comm.
PO Box 4327, 8012 Conser
Overland Park, KS 66204
(913) 381-5900
Receivers, scanners, and accessories

B.C. Communications Inc
The 211 Bldg, Depot Rd
Huntington Station , NY 11746
516-549-8833
Receivers, scanners, and accessories

Barry Electronics
25 Sutton Place So.
New York NY 10012
212-975-7000
CB and ham radios, radio accessories

Burk Electronics
35 N Kensington
La Grange, IL 60525
708-482-9310
Receivers, scanners, and accessories

C. Crane Company
558 10th street
Fortuna, CA 95540-2350
(800) 522-8863
Receivers and accessories

Cable X-perts, Inc.
113 McHenry Rd., Suite 240
Buffalo Grove, IL 60089-1797
(800) 823-3340
Cables and wire

COMMtronics Engineering
PO Box 262478
San Diego, CA 92196
(619) 578-9247
*Scanner/computer interfaces;
modification kits; newsletter; books*

Communications City
175 S E 3rd Ave
Miami, FL 33131
(305) 579-9709
Receivers, scanners, and accessories

Communications Electronics Inc.
PO Box 1045
Ann Arbor, Michigan 48106-1045
(800) 872-7226
Receivers and accessories

Communications Specialists, Inc.
426 West Taft Avenue
Orange, CA 92665-2496
(800) 854-0547
CTCSS tone decoders; related products

Copper Electronics, Inc.
3315 Gilmore Industrial Blvd.
Louisville , KY 40213
(800) 626-6343   Fax 502-968-0449
Citizens Band radios and accessories

Electronic Distributors (EDCO)
325 Mill Street
Vienna, VA 22180
(703) 938-8105
Receivers, scanners and accessories

Electronic Equipment Bank
323 Mill Street NE
Vienna, VA 22180
Receivers, scanners, and accessories

Fair Radio Sales
1016 E. Eureka, Box 1105
Lima, Ohio 45802
(419) 227-6573
Used receivers and military surplus

FM Atlas Publishing
Box 336
Esko, MN 55733-0336
218-879-7676   Fax: 218-879-8333
FM Atlas directories; SCS kits

Gilfer Shortwave
52 Park Avenue
Park Ridge, NJ 07656
(800) 445-3371
Amateur radio transceivers; receivers

Grove Enterprises
PO Box 98
Brasstown, NC 28902
(800) 438-8155
Receivers, scanners, and accessories

Grundig
PO Box 2307
Menlo Park, CA 94026
(800) 872-2228
Shortwave receivers

Ham Radio Outlet
390 Diablo Rd, Ste 210
Danville, CA 94526
Amateur radio transceivers and other
receivers [nationwide local outlets]

Henry Radio Inc.
2050 Bundy Drive
Los Angeles, CA 90025
(800) 877-7979   310-820-1234
Amateur radio transceivers; amplifiers

Icom America, Inc.
2380-116th Ave. N.E.
Bellvue, WA 98004
(206) 454-7619
Amateur radio transceivers; receivers

Japan Radio Co. Ltd.
430 Park Ave (2nd Floor)
New York, NY 10022
(212) 355-1180
Shortwave receivers

Jun's Electronics
5563 Sepulveda Blvd
Culver City, CA 90230
(800) 882-1343
Receivers, scanners and accessories

Kenwood Communications Corp
Box 22745-2201 E. Dominguez St
Long Beach, CA 90801-5745
(310) 639-5300
Amateur radio transceivers; receivers

Kiwa Electronics
612 South 14th Ave.
Yakima, WA 98902
(800) 398-1146
Accessories

Lentini Communications
21 Garfield Street
Newington, CT 06111
Amateur radio transceivers; receivers

Lowe Electronics Ltd.
Chesterfield Road
Matlock, Derbyshire
DE4 5LE, England
Receivers

MFJ Enterprises, Inc.
Box 494
Mississippi State, MS 39762
(800) 647-1800
Accessories

Michigan Radio
23040 Schoenherr
Warren, MI 48089
Amateur radio transceivers; receivers

Optoelectronics
5821 NE 14th Ave.
Fort Lauderdale, FL 33334
(800) 327-5912
Radio accessories

Palomar Engineers
Box 462222
Escondido, CA 92046
(619) 747-3343
Radio accessories

Passport RDI White Papers
Box 300
Penn's Park, PA 18943
(215) 794-8252
Receiver reviews

PerCon Corporation
4906 Maple Springs / Ellery Road
Bemus Point, NY 14712
(716) 386-6015
FCC databases on CD-ROM

Portland Radio Supply
234 Se Grand Ave.
Portland, OR 97214
503-233-4904
Amateur radio transceivers; receivers

R. L. Drake Company
PO Box 3006
Miamisburg, OH 45343
(800) 568-3426
Shortwave receivers

Radio Center USA
12 Glen Carran Circle
Sparks, NV 89431
800 345-5686 702-331-7373
Amateur radio transceivers; receivers

Radio Shack
1500 One Tandy Center
Ft. Worth, TX 76102
[6600 nationwide stores]

Ramsey Electronics, Inc.
793 Canning Parkway
Victor, NY 14564
(800) 446-2295
Accessories and kits

Rosetta Laboratories Pty. Ltd.
222 St. Kilda Road
St. Kilda, Victoria 3182, Australia
Phone: +61 3 9525 5300
Fax: +61 3 9525 3560
http://www.winradio.net.au
E-Mail: info@winradio.net.au
WinRadio on a PC card; freq database

Sangean America, Inc.
2651 Troy Ave.
South El Monte, CA 91733
(818) 579-1600
Receivers

Satellite City
2663 Country Road 1
Minneapolis   MN   55112
(800) 426-289, (612) 786-4475
Amateur radio transceivers; receivers

Scan Communication Co.
PO Box 911
Burlington, IA 52601
Amateur radio transceivers; receivers

Scanner World USA
10 New Scotland Ave
Albany, NY 12208
(518) 436-9606
Scanners and related accessories

Tandy Corporation
300 W. Third Street
Ft. Worth, TX 76102
(817) 390-3569
Receivers, scanners, and accessories

Ten - Tec, Inc.
1185 Dolly Parton Parkway
Sevierville , TN 37862
(800) 833-7373
Amateur radio transceivers; receivers

The Radio Place
5675-A Power Inn Rd
Sacramento, CA   95824
(916) 387-0730
Receivers, scanners, and accessories

U.S. Radio
377 Plaza
Granbury, TX   76048
Receivers, scanners, and accessories

Trucker Electronics
1717 Reserve Street
Garland, TX 75042
(800) 527-4642
Receivers and accessories

Uniden America Corporation
4700 Amon Carter Blvd
Ft. Worth, TX   76155
Receivers, scanners, and accessories

Universal Radio, Inc.
6830 Americana Pkwy
Reynoldsburg, OH 43068
(800) 431-3939
Receivers and accessories

Viking International
150 Executive Park Blvd.,
Suite 4600
San Francisco, CA 94134
(415) 468-2066
Long-play tape recorders

Worldcom Technology
PO Box 3364
Ft. Pierce, FL 34948
(407) 466-4640
Antennas

Yaesu USA
17210 Edwards Rd.
Cerritos, CA 90701
(310) 404-2700
Amateur radio transceivers; receivers

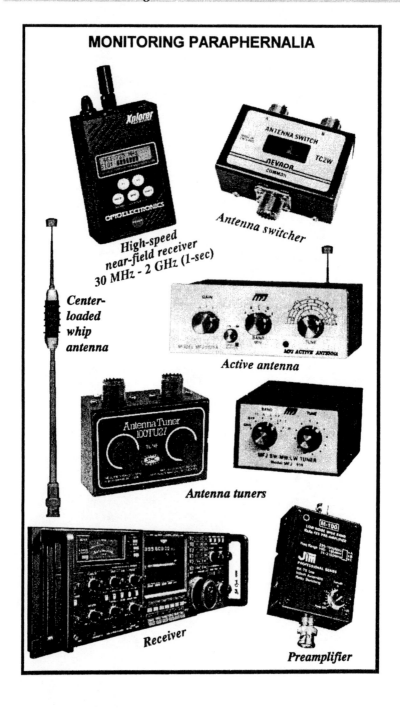

MONITORING PARAPHERNALIA

High-speed near-field receiver 30 MHz - 2 GHz (1-sec)

Antenna switcher

Center-loaded whip antenna

Active antenna

Antenna tuners

Receiver

Preamplifier

Appendix 2

Selected bibliography

Over the last few years I have had the pleasure of writing the "Listener's Library" column for *The Journal of the North American Shortwave Association*. In this role I have had the chance to review dozens of books related to the radio hobby. In addition to the books and publications mentioned throughout the chapters, here is a short list of titles that may be of some help to you as a person starting out in the radio monitoring hobby. This list merely scratches the surface of books available to the radio hobbyist. Remember, knowledge is power! A couple of good radio books can take you a long way. This book did, didn't it?

AMATEUR RADIO ENCYCLOPEDIA
Stan Gibilisco W1GV, Editor in Chief
Tab Books, Blue Ridge Summit, PA
ISBN 0-8306-4096-7

THE ANTENNA HANDBOOK
A Guide to Understanding and
Designing Antenna Systems
by W. Clem Small
Grove Enterprises, Brasstown, NC
ISBN: 0-944543-07-3

THE ARRL ANTENNA HANDBOOK
The American Radio Relay League,
Newington, CT
ISBN 0-87259-473-4

THE ARRL HANDBOOK FOR RADIO
AMATEURS
The American Radio Relay League,
Newington, CT
ISBN 0-87259-171-9

BASIC ELECTRONICS THEORY,
Fourth Edition by Delton T. Horn
Tab Books, Blue Ridge Summit, PA
ISBN: 0-8306-4199-8

THE BEGINNER'S HANDBOOK OF
AMATEUR RADIO
by Clay Laster, W5ZPV
Tab Books, Blue Ridge Summit, PA
ISBN: 0-8306-4354-0

BUILD YOUR OWN SHORTWAVE
ANTENNAS, Second Edition
by Andrew Yoder
Tab Books, Blue Ridge Summit, PA
ISBN 0-07-076534-0

BUYING A USED SHORTWAVE
RECEIVER
Edited by Fred Osterman
Universal Radio Research,
Reynoldsburg, Ohio
ISBN 1-882123-09-3

THE CLANDESTINE
BROADCASTING DIRECTORY
by Mathias Kropf
Tiare Publications, Lake Geneva, WI

COMPUTERIZED RADIO
MONITORING
by Todd D. Dokey
Tiare Publications, Lake Geneva, WI
ISBN: 0-936653-44-2

DICTIONARY OF SCANNER
TERMS, SLANG, AND
ABBREVIATIONS
by Les Mattson
DX Radio Supply, Wagontown, PA

EASY-UP ANTENNAS FOR RADIO
LISTENERS AND HAMS
by Edward M. Noll, W3FQJ
MFJ Enterprises, Inc.,
Mississippi State, MS

EMERGENCY RADIO!
Scanning News As It Happens
by Norm Schrein
Index Publishing Group, San Diego, CA
ISBN: 1-56866-050-2

FERRELL'S CONFIDENTIAL
FREQUENCY LIST
Compiled by Geoff Halligey
Gilfer Associates, Inc., Park Ridge, NJ
ISBN: 0-914542-24-9

JOE CARR'S RECEIVING ANTENNA
HANDBOOK
by Joe Carr K4IPV
HighText Publications Inc., Solana
Beach, CA
1-800-888-4741

THE PIRATE RADIO DIRECTORY
by Andrew Yoder and George Zeller
Tiare Publications, Lake Geneva, WI
ISBN: 0-936653-57-4

PROPAGATION PROGRAMS:
A REVIEW OF CURRENT
FORECASTING SOFTWARE
By Jacques D'Avignon VE3VIA
Grove Enterprises, Inc., Brasstown, NC

RADIO RECEIVER PROJECTS YOU
CAN BUILD by Homer L. Davidson
Tab Books, Blue Ridge Summit, PA
ISBN 0-8306-4190-4

SCANNER MODIFICATION
HANDBOOKS, Vols 1 and 2
by Bill Cheek
CRB Research Books, Inc.,
Commack, NY 11725
Vol. 1 = ISBN 0-939780-11-9
Vol. 2 = ISBN 0-939780-14-3

SCANNER RADIO GUIDE
by Larry M. Barker
HighText Publications Inc.,
Solana Beach, CA
ISBN: 1-878707-10-8

SCANNER MODIFICATIONS AND
ANTENNAS by Jerry Pickard
Index Publishing Group, Inc.,
San Diego, CA
ISBN: 1-56866-120-7

SCANNERS & SECRET
FREQUENCIES
by Henry L. Eisenson
Index Publishing Group, Inc.,
San Diego, CA
ISBN 1-56866-038-3

SHORTWAVE LISTENING
GUIDEBOOK: THE COMPLETE
GUIDE TO HEARING THE WORLD
by Harry Helms
HighText Publications, Inc.
Solana Beach, CA
ISBN: 1-878707-11-6

SHORTWAVE LISTENER'S GUIDE
FOR APARTMENT/CONDO
DWELLERS
by Edward M. Noll, W3FQJ
MFJ Enterprises, Inc.,
Mississippi State, MS

SHORTWAVE RECEIVERS,
PAST & PRESENT
Edited by Fred Osterman
Universal SW Radio Research,
Reynoldsburg, Ohio

TUNING INTO RF SCANNING
FROM POLICE TO SATELLITE
BANDS
By Bob Kay
Tab Books, Blue Ridge Summit, PA
ISBN 0-07-033964-3

THE ULTIMATE SCANNER
(CHEEK 3)
by Bill Cheek
Index Publishing Group,
San Diego, CA
ISBN 1-56866-058-8

THE UNDERGROUND FREQUENCY
GUIDE
A Directory of Unusual, Illegal, and
Covert Radio Communications
by Donald W. Schimmel
HighText Publications Inc., Solana
Beach, CA
ISBN 1-878707-17-5

THE WORLD IS YOURS
ON SHORTWAVE RADIO
by Samuel Alcorn
Gilfer Associates, Inc., Park Ridge, NJ
ISBN: 0-914542-23-6

MONITORING LOG

| Date | Time | Frequency | Mode | Call/ID | Signal Quality | Report | | Comments/Notes | |
	GMT/UTC	kHz or MHz	AM/FM SSB&CW etc		SINPO	Sent	Rcvd		
									1
									2
									3
									4
									5
									6
									7
									8
									9
									10
									11
									12
									13
									14
									15
									16
									17
									18
									19
									20

Notes/Comments/Doodles

The basic log format for monitoring below 30 MHz

Appendix 3

Radio monitoring on-line

If you happen to be a computer-oriented radio monitor, you are probably interested in on-line services that support the radio hobby. All of the major on-line services (America Online, CompuServe, Delphi, GEnie) have message and bulletin areas dedicated to the radio hobby. Most also have file download areas dedicated to the hobby. If you have Internet access, either direct or through an online service, several Internet Newsgroups support our hobby. These are:

    rec.radio.amateur.antenna
    rec.radio.amateur.homebrew
    rec.radio.amateur.space
    rec.radio.scanner
    rec.radio.pirate
    rec.radio.shortwave
    rec.radio.swap

All of these Newsgroups support lively discussions and information exchanges that might be of use to the computer-connected monitor.

If your Internet provider supports FTP (File Transfer Protocol), several file sites have many monitoring related files to choose from.:

> ftp.cs.buffalo.edu
> ftp.demon.co.uk/pub/ham
> ftp.sunset.se/pub/radio
> mgate.arrl.org
> oak.oakland.edu/pub/hamradio

If you do have full Internet access through whatever provider you choose, the World Wide Web is becoming the place to find information for many aspects of the radio world. If you have "web-crawling" capability, give a look at the following home pages:

> http://www.acs.ncsu.edu:80/HamRadio/
> http://amsat.org/amsat/AmsatHome.html
> http://www.acs.oakland.edu/barc.html
> http://w6yx.stanford.edu/w6yx.html
> http://www.waverider.co.uk/~paulj/radio.html
> http://www.panix.com/clay/scanning
> hhtp://www.cs.nmsu.edu/~thharrel/scanner/scanner.html
> http:/www.grove.net/
> http://ourworld.compuserve.com/homepages/bcheek
> http://wcoil.com/~gnbc/
> http://www.pics.com/gilfer/

This is by no means a complete list, but see pages 319 and 320 for many more Web sites in the U.S. and around the world. Internet sites and services come and go. When you connect with a site, check it for a current list of other radio-related computer places where hobbyists gather.

ACE	http://www.access.digexnet/~cps/ACE.html
ANARC	http://www.anarc.org/
ARDXC	http://maya.eagles.bbs.net.au/~ardxc/
BBC WS	http://www.bbcnc.org.uk/worldservice/
Benelux DX Club	http://promet12.cineca.it/htdx/swls/bdxc.html
Bob Colyard's SPEEDX site	http://cybercomm.net/~slapshot/speedx.html
Byron Hicks' Shortwave Radio Schedule Guide	http:// aloha.nmsu.edu/w5gb/swl/swl.html
Chilton's R8	http://www.chilton.com/scripts/radio/R8-receiver
COMMtronics Engineering and The World Scanner Report	http://www.ourworld.compuserve.com/homepages/ bcheek
Deutsche Welle	http://www-dw.gmd.de/
DX Antwerp	http://www.innet.net/~dxa/
DX Club of Irkutsk	http://www.icc.ru/fed/radio.html
EarthView	http://www.fourmilab.ch/cgi-bin/uncgi/Earth/action?opt=p
EEB	http://www.access.digex.net/~eeb/eeb.html
FCC Databases	http://www.radiostation.com/ http://radio.aiss.uiuc.edu~rrb/feedb.html
Gilfer Shortwave	http://www.gilfer.com/
Graffiti Board Log Book	http://espresso.ts.uvic.ca/cgi/log.exe
Grove	http://www.grove.net/
Ham Radio Outlet	http://www.hamradio.com/
Hans Bakhuizen's Broadcasting Corner	http://www.wp.com/hansbakhuizen
IGB U.S. Mirror	http://www.cs.cmu.edu/~jblythe/short-wave.html
Index Publishing Group	http://www.electriciti.com/~ipgbooks
Internet Guide to Broadcasters	http://www.informatik.uni-oldenburg.de/~thkoch
IPS Space Services	http://www.ips.oz.au/
IRRS	http://www.nexus.org/
Java Greyline Map	http://www.infi.net/~dharvey/grayline/grayline.html
JPS Communications	http://emporium.turnpike.net/J/JPS/jps.html
Kenwood	http://www.kenwood.net/
Kiwa	http://www.wolfe.net/~kiwa
Klingenfuss	http://ourworld.compuserve.com/homepages/ klingenfuss/
Lowe	http://www.lowe.co.uk/
Medium Wave Circle	http://www.cs.vu.nl/~gerben/mwc/
NASWA	http://www.mcs.com/~gerben/radio.html
NOAA	http://www.sel.noaa.gov/
Nordic SW Center	http://sunweb.sds.se/org/swl/
NRC	http://alpha.wcoil.com/~gnbc/

Paul Dwerryhouse's ShortWWWave	http://ariel.ucs.unimelb.edu.au/~pbd/SW/index.html
Radio Australia	http://www.abc.net.au/ra/
Radio Austria International	http://www.ping.at/rai/
Radio Canada International	http://radioworks.cbc.ca/radio/rci/rci.html
Radio Japan	http://www.nhk.or.jp.rjnet/
Radio Netherlands	http://www.rnw.nl/
Radio Slovakia International	http://www.xs4all.nl/~xavcom/rozhlas/
Radio Sweden	http://www.sr.se/rs/
Radio Television Malaysia	http://www.asiaconnect.com.my/rtm-net/
Radio Vlaanderen International	http://www.brtn.be/rvi/
Real Audio	http://realaudio.com/
Shortwave Radio Catalog	http://itre.ncsu.edu/radio/
Sites Solar Terrestrial Dispatch	http://solar.uleth.ca/
The FCC	http://www.fcc.gov/
Universal Radio	http://www.universal-radio.com/
Voice of America	gopher://gopher.voa.gov/
WebO'Visions USA Radio Links	http://www.webovision.com/media/sd/usradio.html
World Radio Network	http://www.wrn.org/
World Wide DX Club	http://ourworld.compuserve.com/homepages/wwdxc/
Xing Stream Works	http://www.xingtech.com/
Yaesu	http://www.yaesu.com/

Radio spectrum allocations

The radio frequency spectrum begins just a hair above DC (*0 Hz*) and extends to far beyond light (*infinity*). The useful portion to date is from a few Hertz up to the frequency of x-rays or so (*far beyond the terahertz*). The regulated or allocated portion of this spectrum is only a "sliver," 9 kHz to 300 GHz.

This appendix attempts to emphasize current USA allocations for the **monitorable** 1½% of that spectrum (0–3.5 GHz), but changes are frequent and errors impossible to avoid. So the following Table is to be taken as a guide and supplement to Chapter 4, "*A DC-To-Daylight Frequency Guide.*"

The data comes from official government documents and other reasonably informed sources, and therefore is more formal than what appears in the body of this book. Abbreviated and edited, it is not cosmic-accurate, but I believe you will find it relevant and helpful to your monitoring adventures.

The monitorable RF spectrum

From (MHz)	To (MHz)	Description
0.0000	0.0090	Not Allocated (0–9 kHz)
0.0090	0.0140	Radionavigation (9–14 kHz)
0.0140	0.0195	Fixed/Maritime Mobile
0.0195	0.0205	Standard Frequency/Time
0.0205	0.0590	Fixed/Maritime Mobile
0.0590	0.0610	Standard Frequency/Time (WWVB 60 kHz)
0.0610	0.0700	Fixed/Maritime Mobile
0.0700	0.0900	Fixed/Maritime Mobile/RadioLocation
0.0900	0.1100	Radionavigation (LORAN)
0.1100	0.1300	Fixed/Maritime Mobile/RadioLocation
0.1300	0.1900	Fixed/Maritime Mobile
0.1900	0.2000	Aero Radionavigation
0.2000	0.2850	Aero Radionavigation/Aero Mobile
0.2850	0.3250	Maritime Radionavigation/Beacons/
0.3250	0.3350	Aero Radionavigation/Aero Mobile
0.3350	0.4050	Aero Radionavigation/Beacons/Aero Mobile
0.4050	0.4150	Radionavigation/Aero Mobile
0.4150	0.4350	Aero Radionavigation/Radionavigation
0.4350	0.4950	Maritime Mobile/Aero Radionavigation
0.4950	0.5050	Emergency And Distress
0.5050	0.5100	Maritime Mobile
0.5100	0.5250	Aero Radionavigation
0.5250	0.5350	Aero Radionavigation/Beacons
0.5350	1.7050	AM Broadcast North America
1.7050	1.8000	Fixed Service Land/Mobile/Marine
1.8000	2.0000	Amateur (160-m)
2.0000	2.1070	Maritime Mobile
2.1070	2.1700	Fixed Service Land/Mobile/Marine
2.1700	2.1940	Land Mobile Service
2.1940	2.3000	Fixed Service
2.3000	2.4950	Shortwave Broadcast (120-m)
2.4950	2.5050	Standard Time/Frequency (WWV)
2.5050	2.8500	Fixed; Land Mobile; Marine
2.8500	3.1550	Aero Mobile Transoceanic Flights
3.1550	3.2000	Fixed Service
3.2000	3.4000	Shortwave Broadcast (90-m)
3.4000	3.5000	Aero Mobile Transoceanic Flights
3.5000	4.0000	Amateur (80/75-m)
3.9000	4.0000	Shortwave Broadcast (75-m)
4.0000	4.0000	Standard Time/Frequency
4.0000	4.0630	Fixed Service
4.0630	4.4380	Maritime Mobile Ship / Shore
4.4380	4.6500	Fixed Service
4.6500	4.7500	Aero Mobile Transoceanic Flights
4.7500	4.8500	Fixed; Mobile; no aero mobile
4.8500	4.9950	Fixed; Mobile
4.7500	*5.0600*	*International Shortwave Broadcast (60-m)*
5.0000	5.0000	Standard Time/Frequency (WWV)
5.0050	5.4500	Fixed Service

5.4500	5.7300	Aero Mobile Transoceanic Flights
5.7300	5.9500	Fixed Service
5.9500	6.2000	Shortwave Broadcast (49-m)
6.2000	6.5250	Maritime Mobile Ship / Shore
6.5250	6.7650	Aero Mobile Transoceanic Flights
6.7650	7.0000	Fixed Service
7.0000	7.3000	Amateur (40-m)
7.1000	*7.3500*	*International Shortwave Broadcast 41-M*
	7.3350	*Time Standard CHU Canada*
7.3000	8.1000	Fixed; Mobile
8.0000	*8.0000*	*International Time/Standard Frequency (WARC Allocation)*
8.1000	8.8150	Maritime Mobile Ship / Shore
8.8150	9.0400	Aero Mobile Transoceanic Flights
9.0400	9.5000	Fixed Service
9.5000	9.9000	Shortwave Broadcast (31-m)
9.7750	9.9950	Fixed Service
9.9950	10.0050	Standard Time/Frequency (WWV)
10.0050	10.1000	Aero Mobile Transoceanic Flights
10.1000	10.1500	Amateur (30-m; CW Only)
10.1000	11.1750	Fixed Service
11.1750	11.4000	Aero Mobile Transoceanic Flights
11.4000	11.6500	Fixed Service
11.6500	12.0500	Shortwave Broadcast (25-m)
12.0500	12.3300	Fixed Service
12.3300	13.2000	Maritime Mobile Ship / Shore
13.2000	13.3600	Aero Mobile Transoceanic Flights
13.3600	13.6000	Fixed Service
13.6000	13.8000	Shortwave Broadcast
13.8000	14.0000	Fixed Service
14.0000	14.3500	Amateur (20-m)
14.3500	14.9950	Fixed Service
15.0000	15.0000	Standard Time/Frequency (WWV)
15.0100	15.1000	Aero Mobile Transoceanic Flights
15.1000	15.6000	Shortwave Broadcast (19-m)
15.6000	16.4600	Fixed Service
16.4600	17.3600	Maritime Mobile Ship / Shore
17.3600	17.5500	Fixed Service
17.5500	17.9000	Shortwave Broadcast (16-m)
17.9000	18.0300	Aero Mobile Transoceanic Flights
18.0300	18.7800	Fixed Service
18.0680	18.1680	Amateur (17-m)
18.7800	18.9000	Maritime Mobile Ship / Shore
18.9000	19.6800	Fixed Service
19.6800	19.8000	Maritime Mobile Ship / Shore
19.8000	19.9000	Fixed Service
19.9000	20.0100	Standard Time/Frequency (WWV)
20.0100	21.0000	Fixed Service
21.0000	21.4500	Amateur (15-m)
21.4500	21.8500	Shortwave Broadcast (13-m)
21.8500	22.0000	Aero Mobile
22.0000	22.7200	Maritime Mobile Ship / Shore
22.7200	23.2000	Fixed Service

23.2000	23.3500	Aero Mobile
23.3500	24.8900	Fixed Service
24.8900	24.9900	Amateur (12-m)
24.9900	25.0100	Standard Time/Frequency (WWV)
25.0200	25.0600	Industrial
25.0700	25.2100	Maritime Mobile
25.2100	25.5300	Land Mobile Industrial
25.5300	25.5500	Federal Government
25.5500	25.6700	Federal Government
25.6700	26.1000	Shortwave Broadcast (12m)
26.1000	26.1750	Maritime Mobile
25.8700	26.4700	Broadcast Pickup
26.4800	26.9500	Federal Government
	26.6200	Civil Air Patrol
26.9500	26.9600	International Fixed Service
26.9650	27.4050	Citizens Band Class D (11-m)
27.4300	27.5300	Business
27.7100	27.9000	Forest Products
	27.9000	US Army
28.0000	29.7000	Amateur (10-m)
29.7000	29.8000	Government (Forestry Service); private Land-Mobile
29.8000	29.8900	Fixed Service and aero fixed
29.8900	29.9100	Government
29.9100	30.0000	Fixed Service and aero fixed
30.0000	30.5600	Government (Military)
30.5600	30.8400	Industrial; Misc.
30.8600	31.1400	Motor Carriers - Busses
30.8600	31.9800	Forestry Conservation
30.8800	31.2400	Business
31.2800	32.0000	Special Industrial
32.0000	33.0000	Federal Government
33.0000	34.0000	Land Mobile
33.0000	*33.0100*	*Land Transportation/Highway Maintenance/Emergencies*
33.0100	*33.1100*	*Public Safety*
33.1100	*33.4100*	*Business/Industrial (Mostly Petroleum)*
33.4100	*34.0000*	*Public Safety/Fire*
34.0000	35.0000	Federal Government
35.0000	36.0000	Land Mobile
35.0000	*35.1900*	*Business/Industrial*
35.1900	*35.6900*	*Domestic/Mobile Phones (Base)/Paging*
35.6400	*35.6800*	*Public Safety/Paging*
35.6900	*36.0000*	*Special Industrial*
36.0000	37.0000	Federal Government
37.0000	37.5000	Land Mobile
37.0000	*37.0100*	*Business/Industrial*
37.0100	*37.4300*	*Public Safety; Police & Local Government*
37.4300	*37.8900*	*Business/Industrial*
37.8900	*38.0000*	*Public Safety; Police & Local Government*
38.0000	38.2500	Radioastronomy
38.2500	39.0000	Federal Government (Military)
39.0000	40.0000	Public Safety
40.0000	42.0000	Federal Government

42.0000	46.6000	Land Mobile
42.0000	*42.9500*	*Public Safety*
42.9500	*43.1900*	*Business/Special Industrial*
43.1900	*43.6900*	*Domestic/Mobile Phones/Paging/Public Safety*
43.7000	*43.8400*	*Busses*
43.8400	*44.4400*	*Trucks*
44.4400	*44.6000*	*Busses*
44.6000	*45.0400*	*Forestry*
44.6100	*46.6000*	*Public Safety*
46.6000	47.0000	Federal Government
46.6100	*46.9700*	*Cordless Phones (Bases)*
47.0000	49.6000	Land Mobile
47.0000	*47.4300*	*Public Safety*
47.4300	*47.6900*	*Public Safety/Industrial/Special Emergency (Red Cross: 47.420)*
47.6900	*48.5400*	*Power and Water Utilities*
48.5400	*49.6000*	*Petroleum/Forestry*
49.6000	50.0000	Federal Government
49.6700	*49.9900*	*Cordless Phones (Handsets); Walkie Talkies*
50.0000	54.0000	Amateur Radio (6-m)
54.0000	72.0000	VHF Television (Channels 2-4)
72.0000	73.0000	Fixed; Mobile: Radio Control/Industry/Callboxes/Paging
73.0000	74.6000	Radioastronomy
74.6000	74.8000	Fixed; Mobile; Auditory Assistance Devices
74.8000	75.2000	Aeronautical RadioNavigation
	75.0000	Aero Radionavigation Marker Beacon
75.2000	76.0000	Fixed; Mobile: Radio Control/Industry/Callboxes/Paging
76.0000	88.0000	VHF Television (Channels 5-6)
88.0000	107.9000	FM Commercial Broadcasting (200 kHz spacing)
108.0000	117.9750	Aviation Navigation (VOR)
117.9750	121.9375	Aero Mobile; Air Traffic Control
	121.5000	Emergency Search And Rescue; ELT
121.9375	123.0875	Airport Utility/Instruction/Advisories/Private aircraft
123.0875	123.5875	Aero mobile; Unicom/Multicom
	123.1000	Search & Rescue - Scene of action
	123.6000	FSS Air Carrier Advisory
123.5875	137.0000	Aero mobile; Instruction/Flight Test Control (Mil; Civ; ARINC)
137.0000	138.0000	Space Ops/Weather/Sat/Research/Satellite Comms
138.0000	144.0000	Government (Military; Research)
	143.7500	Civil Air Patrol
144.0000	148.0000	Amateur Radio (2-m)
148.0000	149.9000	Experimental/Business/Satellites/Gov't
149.9000	150.0500	Radionavigation/Satellite/*Russian Military Satellites*
150.0500	150.800	Government
150.8000	156.2475	Land Mobile
150.8000	*150.9800*	*Transportation*
150.9800	*151.4825*	*Public Safety*
	150.9800	*Oil Spill Cleanup*
151.4825	*151.4975*	*Industrial*
151.4975	*152.0000*	*Industrial/Public Safety*
152.0000	*152.2550*	*Domestic/Public*
	152.0075	*Medical Paging*
152.2550	*152.4650*	*Transportation*

152.46500	*152.4950*	*Special Industrial*
152.8850	*153.7325*	*Domestic/Public*
153.7325	*154.4825*	*Industrial/Public Safety*
	153.8300	*Fire - On-Scene Fire fighting*
	154.2000	*Earth Telecommand*
154.4825	*156.2475*	*Industrial*
156.2475	157.4500	Maritime Mobile; Government and private
157.4500	161.5750	Land Mobile
157.4500	*157.7250*	*Transportation*
157.7250	*157.7550*	*Industrial: Forestry/Petroleum/Manufacturers*
157.7550	*158.1150*	*Domestic/Public*
158.1150	*158.4750*	*Industrial*
158.7150	*159.4800*	*Public Safety*
159.4800	*161.5750*	*Transportation*
160.2150	*161.5650*	*Railroads*
161.5750	161.6250	Maritime Mobile
	161.5800	Oil Spill Cleanup
	161.6000	Marine - Port Operations
161.6250	161.7750	Land Mobile: Broadcast Pickup (Radio/TV)
161.7750	162.0125	Maritime Mobile
162.0125	173.2000	Federal Government
	173.0750	Stolen Vehicle Recovery Systems
173.2000	173.4000	Industrial; Public Safety: Newspapers/Motion Pictures
173.4000	174.0000	Government
174.0000	216.0000	VHF Television (Ch 7-13)
216.0000	220.0000	Maritime Mobile; Aeronautical mobile; Fixed; Land Mobile
218.0000	*218.5000*	*Interactive Video Data Service*
220.0000	222.0000	Land Mobile; Radiolocation/Radar
222.0000	225.0000	Amateur Radio (1.3-m); Radar
225.0000	328.6000	Government (Mil Aviation; MilSat; fixed; mobile)
	243.0000	Government (Mil Aviation; Emergency; ELT)
328.6000	335.4000	Government (Glide Slope; AeroNavigation)
335.4000	399.9900	Government (Mil Aviation; Emergencies 243.0 and 282.8)
399.9900	400.0500	RadioNavigation; mobile-satellite
400.0500	400.1500	Standard Time & Frequency Signal-Satellite
400.1500	406.1000	Weather aids; weather satellite; space communication
406.1000	410.0000	Radio astronomy; Government
410.0000	420.0000	Government (All Agencies)
420.0000	450.0000	Amateur Radio (70-cm); radar
450.0000	470.0000	Land Mobile
450.0000	*451.0000*	*Broadcast Pickups (Radio and TV)*
451.0000	*454.0000*	*Public Safety/Industrial/Transportation*
454.0000	*455.0000*	*Domestic/Public*
455.0000	*456.0000*	*Broadcast Pickups (Radio and TV)*
456.0000	*459.0000*	*Public Safety/Industrial/Transportation*
459.0000	*460.0000*	*Domestic/Public*
460.0000	*462.5375*	*Public Safety/Industrial/Transportation*
462.5375	*462.7375*	*General Mobile Radio Service (GMRS)*
462.9375	*463.1875*	*Medical (Med1-Med8;Dispatch)*
462.7375	*467.5375*	*Public Safety/Industrial/Transportation*
467.5375	*467.7375*	*General Mobile Radio Service (GMRS)*
467.9375	*468.1875*	*Medical (Med1-Med8;Dispatch)*

467.7375	*470.0000*	*Public Safety/Industrial/Transportation*
470.0000	806.0000	UHF Television (Ch 14 - 70)
470.0000	*512.0000*	*Public Safety/Local Government in some areas (ex: Los Angeles)*
608.0000	*614.0000*	*Radioastronomy*
806.0000	902.0000	Land Mobile
806.0000	*821.0000*	*Conventional and trunked (Mobiles)*
821.0000	*823.9500*	*Public Safety (Mobiles)*
823.9500	*849.0000*	*Cellular Mobile Telephone (Mobiles)*
849.0000	*851.0000*	*Reserved for Expansion*
851.0000	*866.0000*	*Conventional and trunked (Bases)*
866.0000	*868.9500*	*Public Safety Service (Bases)*
868.9500	*894.0000*	*Cellular Mobile Telephone (Bases)*
894.0000	*896.0000*	*Reserved for Expansion*
896.0000	*901.0000*	*Private Land Mobile Radio Service*
901.0000	*902.0000*	*Reserved for Expansion*
902.0000	928.0000	Amateur Radio (33-cm)
902.0000	928.0000	Radiolocation/radar; ISM; *Cordless Phones*
928.0000	929.0000	Fixed
929.0000	932.0000	Land Mobile
932.0000	935.0000	Government/Private Fixed
935.0000	940.0000	Land Mobile
940.0000	941.0000	Mobile
941.0000	944.0000	Government/Private; Fixed
944.0000	960.0000	Private; fixed
944.0000	*952.0000*	*Broadcast Radio Service (STL; Relays)*
952.0000	*956.1000*	*Private; Fixed; Microwave Service*
956.2620	*956.4375*	*Private; Fixed; (Signaling And Control)*
956.5000	*959.8000*	*Private; Fixed; Microwave Service*
959.8620	*960.0000*	*Common Carrier Radio Service*
960.0000	1215.0000	Aviation Services (Tacan; DME) 1090.0 Air-to-Ground
1215.0000	1240.0000	Radar/Radionavigation
1240.0000	1300.0000	Amateur; radar
1300.0000	1350.0000	Aeronautical radionavigation
1350.0000	1400.0000	Government; Fixed-Mobile; Radar
1400.0000	1427.0000	Earth Exploration-Satellite-Radioastronomy Space Research
1427.0000	1429.0000	Fixed; Mobile; Space Operations; Telemetry
1429.0000	1435.0000	Fixed; Mobile; Land Mobile; Telecommand; Telemetry
1435.0000	1530.0000	Mobile; Aerotelemetry
1530.0000	1535.0000	Maritime-Mobile-Satellite Aerotelemetry
1535.0000	1545.0000	Maritime-Mobile-Satellite
1545.0000	1559.0000	Aero Mobile-Satellite
1559.0000	1610.0000	Aero Radionavigation Satellite
1610.0000	1626.5000	Aero Radionavigation Radiodetermination Mobile-Satellite
1626.5000	1645.5000	Maritime-Mobile-Satellite (Up)
1645.5000	1646.5000	Mobile-Satellite (Up)
1646.5000	1660.0000	Aero Mobile-Satellite (Up)
1660.0000	1660.5000	Aero Mobile-Satellite (Up) Radioastronomy
1660.5000	1668.4000	Radioastronomy Space Research
1668.4000	1670.0000	Weather Aids Radioastronomy
1670.0000	1700.0000	Weather Aids Weather Satellite (Down)
1700.0000	1710.0000	Fixed Weather Satellite (Up)
1710.0000	1850.0000	Government

1850.0000	2150.0000	Fixed-Mobile
2150.0000	2160.0000	Fixed
2160.0000	2200.0000	Fixed-Mobile
2200.0000	2290.0000	Government
2290.0000	2300.0000	Fixed; Mobile; Deep Space Research
2300.0000	2310.0000	Amateur Band (13-cm)
2310.0000	2360.0000	Broadcasting Satellite Mobile
2360.0000	2390.0000	Mobile; Radar
2390.0000	2450.0000	Amateur Band (12-cm)
2450.0000	2483.5000	Fixed; Mobile; Radar
2483.5000	2500.0000	Radiodetermination Satellite (Down)
2500.0000	2655.0000	Broadcasting-Satellite Fixed
2655.0000	2690.0000	Broadcast-Satellite Fixed Earth Exploration Radioastronomy Space
2690.0000	2700.0000	Earth Exploration-Satellite-Radioastronomy Space Research
2700.0000	2900.0000	Government
2900.0000	3100.0000	Maritime Radionavigation
3100.0000	3300.0000	Radar
3300.0000	3500.0000	Amateur Radiolocation

Abbreviations and definitions

Aero:	aeronautical or aviation
ARINC:	Aeronautical Radio, Inc.
DME:	distance measuring equipment
Down:	downlinks; satellite to earth
Fixed:	non-mobile, fixed location transmissions
ISM:	Industrial-Scientific-Medical
Radar:	radio detection and ranging
Radioastronomy:	typically passive; for listening only
Radiolocation:	typically radar and beacons
Radionavigation:	signals that react with remote equipment for navigation
STL:	studio-transmitter links used by broadcasters
Tacan:	tactical air navigation (aero equipment)
Telecommand:	control of remote equipment by radio
Telemetry:	usually non-voice data
Up:	uplinks; earth to satellite
VOR:	VHF omnidirectional range (aero equipmentt)

Appendix 5

Biographical statement

Thomas J. "Skip" Arey, WB2GHA, began his radio monitoring career in the fourth grade by connecting a fifty foot longwire to a five-tube "superhet" AM radio. His lifelong love of radio, coupled with his ability to spin a yarn and teach people a thing or two, led to his becoming a radio hobby journalist.

He has written the "Beginner's Corner" column for *Monitoring Times* magazine since 1989. His articles on the radio hobby have appeared in *American Scannergram, North East Scanning News, The Journal of the North American Shortwave Association,* and *DX News.* Reprints of some of his articles have become standard tools for teaching people about the radio hobby. Skip is a Life Member of The American Radio Relay League. He holds an Extra Class

Amateur Radio License WB2GHA and a commercial General Radio Operators License. He has led lectures and workshops for beginning radio enthusiasts throughout the United States. He was recently elected to the Executive Board of the Association of North American Radio Clubs (ANARC).

Skip has shown up on the other side of many monitor's receivers, at various times, as an officer in the United States Army (including service during Operation Desert Storm), as a member of the Amateur Radio Emergency Service (ARES), the Radio Amateur Civil Emergency Service (RACES), and as an Emergency Medical Technician (EMT).

Skip maintains a "Suburban Guerrilla" existence in Southern New Jersey (it's the Electric Blue stucco house with the Pink Flamingos out front, you can't miss it) with his wife Regina and sons Brendan and Patrick. When he is not playing radio or writing about radio he enjoys collecting and reading the works of Robert A. Heinlein.

Index